Beyond Belief

Skepticism, science and the paranormal

Whether ghosts, astrology or ESP, up to 80 per cent of the population believes in one or more aspects of the paranormal. Such beliefs are entertaining, and it is tempting to think of them as harmless. However, there is mounting evidence that paranormal beliefs can be dangerous – cases of children dying because parents favoured alternative remedies over conventional medicine, and 'psychics' who trade on the grief of the bereaved for personal profit and gain. Expenditure on the paranormal runs into billions of dollars each year.

In *Beyond Belief: Skepticism, science and the paranormal* Martin Bridgstock provides an integrated understanding of what an evidence-based approach to the paranormal – a skeptical approach – involves, and why it is necessary. Bridgstock does not set out to show that all paranormal claims are necessarily false, but he does suggest that we all need the analytical ability and critical thinking skills to seek and assess the evidence for paranormal claims.

Martin Bridgstock is Senior Lecturer in the School of Biomolecular and Physical Sciences at Griffith University, Queensland.

Beyond Belief

Skepticism, science and the paranormal

Martin Bridgstock

CAMBRIDGE
UNIVERSITY PRESS

CAMBRIDGE
UNIVERSITY PRESS

University Printing House, Cambridge CB2 8BS, United Kingdom

One Liberty Plaza, 20th Floor, New York, NY 10006, USA

477 Williamstown Road, Port Melbourne, VIC 3207, Australia

314-321, 3rd Floor, Plot 3, Splendor Forum, Jasola District Centre, New Delhi - 110025, India

79 Anson Road, #06-04/06, Singapore 079906

Cambridge University Press is part of the University of Cambridge.

It furthers the University's mission by disseminating knowledge in the pursuit of
education, learning and research at the highest international levels of excellence.

www.cambridge.org
Information on this title: www.cambridge.org/9780521758932

© Martin Bridgstock 2009

Cover design by Marc Martin, Small and Quiet Design
Typeset by Aptara Corp.

First published 2009

A catalogue record for this publication is available from the British Library

Library of Congress Cataloging in Publication data
Bridgstock, Martin.
Beyond belief skepticism, science and the paranormal / Martin Bridgstock.
9780521758932 (pbk.)
Includes Index.
Bibliography
Parapsychology.
Parapsychology and science.
Skepticism.
001.9

ISBN 978-0-521-75893-2 Paperback

*For
my parents,
Connie and Roy,
with love and gratitude*

Contents

Preface

THIS BOOK HAS its genesis in a lengthy period I spent in hospital in 2001. The illness was painful – and I had a near-death experience – but as I recovered, I spent weeks reviewing my own life and the inevitable coming of my own demise. So, I asked myself, what did I want to do with the time that was left? Experiences of this kind are common among middle-aged men, and can lead to monumentally foolish decisions. I claim no special wisdom, but my change of course seems to have been completely beneficial.

I had acquired a great deal of respect for the skeptical movement, especially in the way it had helped stop the creation scientists from infiltrating education in my home state of Queensland, and so I conceived the idea of a course in skepticism at Griffith University. On being released from hospital I began to plan such a course. To my surprise, no one at the university objected to such a radical change in my teaching. When it was taught, the course was a great success. Numbers rapidly rose to unprecedented heights and the enthusiasm and involvement of the students was greater than I had ever witnessed.

There were some reasonable books in the area, I found, but none which quite fitted the course. I decided to do things the hard way and write my own. I was greatly helped in this by study leave from Griffith University in 2005. I spent part of this at the Centre for the Study of Social Change at Queensland University of Technology and the rest at the Centre for Applied Philosophy and Public Ethics at the University of Melbourne. I am grateful to all these institutions.

The Introduction contains data from a survey obtained by the Australian Skeptics Science and Education Foundation from the Queensland Social

Survey at Central Queensland University (Hanley & Mummery 2008). I am grateful to the Foundation for their assistance and to the Survey staff for their efficiency and helpfulness. The Cambridge University Press staff were enthusiastic and efficient, and I also appreciated the perceptive comments of the two anonymous reviewers.

Writing the book itself was reasonably easy: I had developed the main ideas in the course, and putting them on paper was a logical next step. Re-writing the book, making the ideas clear, avoiding repetition and self-contradiction, was much harder, and took years. My beloved family, my wife Vicki and daughters Sundari and Suji, have been understanding of what is needed, and I am grateful to them, in addition to all else. I also appreciate John Elliott's assistance regarding publication.

I must also thank the hundreds of students I have faced in the lecture theatre. The mass of waving hands raising points, the questions in the intervals, the excited emails have all convinced me that universities can and should be much more than efficient factories for training the workforce. The academic fire can still be passed from generation to generation, and I hope this book will play a part.

This book is primarily a work of advocacy. It outlines a method of thinking and argues that humanity would be better off if all of us adopted this method. It also strongly and explicitly links skeptical thinking to ethical thinking. I do believe the book has a number of novel features as well. Of course, the data in the Introduction are completely new. While the model of science outlined in the early chapters is implied in any number of philosophical texts, to my knowledge, the material has never been explicitly presented like this. The strong and explicit link between skepticism and ethics is also novel, though other writers have hinted at it. Nor, to my knowledge, has anyone placed the three principles (burden of proof, Occam's razor and Sagan's balance) at the heart of skepticism, as I have done here.

Some readers may be irritated by my spelling of 'skepticism'. I am following the precedent set by the Australian Skeptics for several decades, and also drawing attention to my focus upon the modern skeptical movement, which is primarily concerned with the investigation of paranormal claims. The uncommon spelling distinguishes modern skepticism from a somewhat different tradition reaching back thousands of years.

My innovations may be wrong, and I am happy to receive criticism. I do believe, however, that the book presents a clear and cogent case for the adoption of skeptical ways of thinking, and also spells out exactly how enjoyable it is. I hope that readers come to the same conclusion.

Martin Bridgstock
School of Biomolecular and Physical Sciences
Griffith University
2009

Introduction
The paranormal and why it matters

LET US BEGIN with a few stories. All are rather startling, and all make a point crucial for this book. In the UK a few years ago four jurors were involved in the trial of a man for murder. One weekend they became drunk, and decided to use an Ouija board. They would contact the spirits of the murder victim and ask who had committed the crime. In the light of the spirits' replies, the jurors had no hesitation in finding the defendant guilty. The Court of Appeal was not impressed with this method of deliberation, and threw out the verdict (*Sydney Morning Herald* 2001).

Next, we will look at the deaths of two young boys in New Zealand. According to prominent medical authorities, both could probably have been saved. Caleb Moorhead was six months old when he died. His parents, Deborah Ann Moorhead and Roby Jan Moorhead were Seventh Day Adventists and also vegans (which meant that they did not eat meat, fish or dairy products). Caleb was suffering from a vitamin B12 deficiency. The parents received many warnings about what would happen, but stuck to their principles – Caleb's mother said that conventional medicine was 'Satan's way' (*Guardian* 2002) and the parents did not change their son's diet. Caleb died in March 2001 of bronchopneumonia, caused by the lack of an easily injected vitamin (Stickley 2002). According to doctors, if Caleb had received a vitamin injection as little as half an hour before his death, he would have survived.

The case of Liam Williams-Holloway generated far more publicity. At the age of three, Liam was diagnosed with neuroblastoma of the jaw – a dangerous and potentially lethal cancer. He was admitted to hospital on New Zealand's South Island. He had two courses of chemotherapy before his parents, Trina Williams and Brendan Holloway, withdrew him from treatment and took him to Gerard and Dawn Uys who practice an alternative therapy called a 'Quantum Blaster' (Hills 2000). The doctors were horrified, arguing that he had a good chance of survival if given medical treatment, and sought a court order to have the child made a ward of the state (*New Zealand Herald* 2000). The family went into hiding. Later, the court order was lifted. Liam was taken to Mexico where an alternative clinic gave him more Quantum Blaster treatment. Liam subsequently died in Mexico. His body was cremated before a full autopsy could be carried out (Hyde 2001).

There are some significant similarities and differences between these two cases. In both cases the parents appeared to believe sincerely that their conduct was right. However, in both cases, what they were doing was strongly condemned by the medical profession. According to oncologists, if Liam had received a full course of chemotherapy, he would have had at least a 50 per cent chance of recovery, and perhaps more (Hyde 2001) And as we have seen, Caleb could have been saved far more easily. In both cases the treatments that the parents chose were unacceptable to modern medicine and science. For example there has never been any evidence of the value of the Quantum Blaster device. It was invented by Royal Raymond Rife, a Californian pathologist and amateur inventor who believed that all diseases – including cancer and dandruff – were caused by organisms that vibrated with disease-specific frequencies (Hills 2000, p. 25). An Australian magazine, *Electronics Australia,* examined the device, and found that all of the components in it were valued at about fifteen dollars. Further, they found that it generated a tiny current which could not even penetrate the skin, let alone destroy micro-organisms allegedly causing cancer (Hills 2000, p. 25).

There were some striking differences in the way the parents were treated after their sons died. The Moorheads were put on trial in 2002 for causing their son's death by failing to provide the necessities of life. They were convicted and sentenced to five years in prison. By contrast, as they went underground, Williams and Holloway received widespread support for their stance, with vigils in support being held throughout New Zealand. No charges were laid, even after the death of their son in Mexico. Indeed, alternative practitioners to this day continue to argue in favour of the correctness of what the parents were doing (e.g. Holden 2006).

Another case, not connected to health, was detailed in a recent issue of the *Skeptical Inquirer.* A woman in Texas, going to a fortune teller for a reading, was told that a curse had been placed on a young man she was attracted to, and that unless the young woman took prompt action, the young man would die of cancer. The fortune teller offered to help lift the curse in exchange for payment. The woman, who had been expecting to pay about $US35 for the advice, eventually parted with over $US25 000. The fortune teller was eventually arrested for this and other fraudulent activities (Davis 2005).

In London, researchers Richard Wiseman and Emma Greening trained actors to visit psychics and tell them their troubles (all of which were invented). Not only did the psychics completely fail to detect that the stories were bogus, they all proceeded to offer to assist with the problems – at prices ranging from £450 to £900 (about $AUD1125 to $AUD2250) (Wiseman & Greening 1998). One wonders how many distressed people have paid such sums.

Finally, the Australian *Skeptic* magazine recently reported the case of a woman who became addicted to psychic hotlines. She ran up a debt of $AUD80 000 accessing these services. To make matters worse, she then took to crime to pay off her debts (Australian Skeptics 2007, p. 6).

What can we say about these examples? Two things are common to all of them. First, they all involved human suffering or injustice. Children died, women were terrorised by threats of curses or ran up huge debts, and a man was convicted of a crime on the doubtful evidence of an Ouija board. The second feature that they all have in common is that they involved belief in paranormal influences and powers. We will discuss the meaning of 'paranormal' later. For now, we should simply note that all the examples involved an element of the strange and the supernatural. Indeed, without belief in the supernatural, none of these events could have occurred.

We might infer from this that the people involved should not have believed in the paranormal, but this may be too harsh. Maybe the Ouija board really could communicate with the spirits of murdered people. Maybe the unfortunate young woman really was the victim of a terrible curse. We are not entitled to sweep aside the possibility without some investigation.

What we can infer, and what seems completely reasonable, is that these events might not have turned out so badly if the people concerned had been able to weigh up the pros and cons of their paranormal beliefs. For example, if the jurors had some information about the reliability of Ouija boards as a source of information, or the Australian woman had had some idea about how useful psychic hotlines really are, perhaps they would not have made the mistakes they did. Given proper information, it is likely that people will reach wiser decisions than they would make in ignorance.

This is the main goal of this book: powerful techniques exist, and these can be learned easily to enable anyone to weigh up the evidence for believing in paranormal claims. We will discover what these techniques are, and how they work.

You may have an objection at this point. 'Does it matter?', you may ask. 'Is the paranormal sufficiently important to warrant this type of investigation?' To answer this, let us look at some statistics. In 2008, Kylie Sturgess and I arranged for some questions to be posed to a cross-section of Queenslanders in a social survey. A cross-section of people were asked to indicate their belief, or lack of belief, in a range of propositions. Kylie Sturgess and I analysed the survey responses, and obtained the results shown in Table Intro.1.

Table Intro.1 Percentage of Queenslanders believing in a range of paranormal propositions, 2008 (number surveyed = 1243)

Belief	Believe or Strongly Believe (%)
Psychic or spiritual healing or the power of the human mind to heal the body	58.6
Creationism, which is the idea that God created human beings pretty much in their present form at one time within the last 10 000 years	37.7
Ghosts, or that spirits of dead people can come back in some places and situations	35.9
That people can hear from or communicate mentally with someone who has died	29.2
That extraterrestrial beings have visited Earth at some time in the past	29.2
Astrology, or that the position of the stars and planets can affect people's lives	28.4

Note that an actual majority of people believe in psychic or spiritual healing, which is worrying in the light of the cases we looked at above, and that significant minorities believe in each of the other propositions. It looks as if beliefs of this kind are quite widespread, and that their adherents, in Australia alone, can be numbered in the millions.

It might be thought that perhaps Queenslanders are more prone to strange beliefs than others. However, as Table Intro.2 shows, this is not so. Comparing the two tables, it looks as if Queenslanders' views resemble those of Australians generally, and the beliefs of both are quite similar to those of the Americans.

Table Intro.2 Belief in selected aspects of the paranormal in Australia and the US

Belief in Australia/USA	Australia 1997 %	USA 2001 %
Ghosts/ghosts or spirits of dead people can come back in some places and situations	40	38
Astrology/Astrology or that the position of the stars and planets can affect people's lives	28	28
Past lives/future lives/reincarnation, that is the rebirth of the soul in a new body after death	30/34	25
Alien visitors (ancient)/Extraterrestrial beings have visited Earth at some time in the past	32	33
Mind reading/Telepathy or communication between minds without using the traditional five senses	36	36
Psychic healing/psychic or spiritual healing or the power of the human mind to heal the body	68	54

(Bridgstock 2003, p. 7)

British polls of paranormal belief (Fox 2004, p. 356) show similar patterns, as do surveys from other parts of the world. We should note that the percentages vary somewhat from poll to poll. This is inevitable. As social scientists know well, asking slightly different questions can often lead to great disparities in the answers. However, the main point seems to be clear. Belief in the paranormal is common and widespread, and there is also disturbing evidence that in some circumstances this belief can be harmful. In short, it looks as if being able to evaluate paranormal claims can save us from assorted forms of unpleasantness should the claims turn out to be false. Widespread ability to evaluate claims in this way might save a great deal of suffering.

This viewpoint acquires more force when we learn that the paranormal is not some fringe activity, involving few people and little money. In fact, it is an enormous industry. Years ago, George O Abell (1981) estimated that billions of dollars a year are spent on astrology alone in the US. In today's money, that would be far more. A recent Australian survey estimated that in 2004 Australians spent about \$AUD1.8 billion on alternative and complementary health care (MacLennan et al 2006). Add to that the (probably) large amounts spent on clairvoyants, UFO research and communicating with the dead, and it is clear that immense sums of money are involved. And, of course, where such sums are involved, it is important to have some way of working out whether we are getting value for money.

AND WHAT IF THE PARANORMAL IS TRUE . . . ?

So far, the reader might have been irritated by the implication that the claims made by paranormalists are false, and that they are harmful. However, the arguments for carefully evaluating paranormal claims apply just as strongly if the claims are true or if the benefits are real.

As we shall see, paranormal claims are so diverse – and in some cases so inconsistent – that they cannot all be true. However what if some of them are true? What if ESP or clairvoyance or creation science can be shown to be justified? The short answer is that our world would never be the same again. Science provides a useful parallel for thinking about this. A few centuries ago, science was the pastime of a handful of people, and in general had little impact upon the lives of most people. Now, it matters terribly.

Why does science matter? There are two answers to this question (Bridgstock et al 1998b). One answer is that science can affect our way of looking at the universe. Think of Galileo, showing that Earth and humanity are not at the centre of the universe, but a small part in constant motion (Morphet 1977). Think of Darwin, showing that humans were not specially created, but part of an ever-changing pageant of life on this planet (Toulmin and Goodfield 1966, pp. 224–31). Less well-known, think of Charles Lyell (a good friend of Darwin's, incidentally), showing that the geology of the earth could be explained by endless amounts of time, and the processes we can see shaping the world today (Toulmin and Goodfield 1966, pp. 167–70). The way we look at ourselves and the world has never been the same since.

Another way in which science is important is through technology. Again and again, scientific breakthroughs have been followed – sometimes quickly, sometimes with long delays – by technological changes which have affected our lives. The most obvious example is the modern computer. It has shaped, and is shaping, our lives most dramatically, yet it owes its origins to a scientific breakthrough by Bardeen and Shockley in the Bell Telephone laboratories in 1948 (Weber 1996). Another example concerns the life-saving antibiotic drug, penicillin, discovered in 1928 by chance by Alexander Fleming, a researcher in London (Ryan 1992). In short, where we acquire knowledge of the universe, we often also acquire the ability to change the universe, and to improve our own place in it. Modern computers, antibiotic drugs, antisepsis, anaesthetics and much more all came about because of science. Thus, science is important because it is changing our lives through technological advance.

If paranormal beliefs turned out to be true, they would also change our lives in the same way. Consider, for example, clairvoyance. Imagine that the ability to see things beyond our normal senses was established. First, of course, it would change our view of ourselves and the world. We would no longer simply be creatures of flesh with limited understanding of the world beyond us. Instead, we would be spiritual beings, able to range beyond our bodies and to unlock the universe's secrets. Thus, clairvoyance, if true, transforms our view of ourselves and the universe. What is more, it is highly likely that clairvoyance could also yield enormous changes. If it could be used reliably, it would make it impossible to get away with murder or any serious crime. It would make it impossible for rogue states to stockpile nuclear weapons, and it might also make it impossible for people to deceive each other. Clearly, clairvoyance could transform the world if it were true, and it is important for us to know whether it is or not. Exactly the same argument applies to all paranormal claims: they are all potentially important to humanity.

In summary, I believe that the paranormal is important, and I argue that all people should regard it as so. A high proportion of the population has some belief in the paranormal and, as we have seen, this leads to the spending of billions of dollars on paranormal-related services. It can lead to the loss by some people of money they cannot afford; in some cases, it has caused the death of young children. In short, if any paranormal claims are believed and false, the consequences can be bad.

Equally, if any paranormal claims are true, then the potentialities are enormous. Our entire view of ourselves and the universe would be radically transformed, and there is always the possibility of turning some of the new knowledge to good use, such as eliminating some diseases or averting terrible wars. In the same way, if paranormal beliefs, regardless of their truth, are beneficial then we should be able to demonstrate this through the use of evidence.

This is the theme of the book. It is not aimed at debunking or destroying paranormal beliefs or claims; it is aimed at showing how rational evaluations can be made of paranormal claims, so that better decisions can be made about them. It sounds like a modest goal, but the effect can be profound. People often find their entire viewpoint changes as they acquire these new intellectual tools.

TWO IMPORTANT POINTS

Two possible misunderstandings should be clarified at the outset. First, it would be easy to assume that this book is a trip into the wonderful world of the supernatural, where wizards, telepathy and communication with the dead are all around. Let us be clear: the book is not any kind of a guided tour of the paranormal. We will not be moving through amazing claims and exclaiming at the mystery of each one. This can be done in the supernatural section of many bookshops, and no additions are needed. We will be looking at assorted paranormal claims as part of the journey to understanding the skeptical viewpoint. At the same time, this book is not a skeptical polemic. You will not be bombarded with the wonders of skepticism and the horrors of the paranormal. The tools of skepticism are introduced, their uses demonstrated and commended, and the rest is up to the reader.

This sounds very liberal, in that the reader decides, at the end, what should be done. But, there is a big catch in this approach, and that is the second point. For several years I have been teaching a course 'Skepticism, Science and the Paranormal' (Bridgstock 2004). The intellectual approaches have always been presented as tools, to be used or not as the students wish. However, the tools are not like chisels or spades. They are powerful intellectual perspectives. Once learned and understood, they become an integral part of a person's outlook. Even if there is a desire to return to the previous state of willing belief, it may be impossible because an awareness has been planted that other, more critical approaches do exist. Thus, in a very literal sense, the approach taken here can be life-transforming and, for some people, profoundly disturbing.

Like any set of tools, skeptical tools work better the more they are used. And there is always the option, at least in principle, of simply leaving them in the box. Still, by the end of this book the reader will know what skepticism is, how to think skeptically, and why this is a good idea.

HOW THE BOOK WORKS

Chapter one might strike the readers as irrelevant. In fact it is pivotal to everything else. We will talk about the nature of science. Why do we have to do this? The answer is that both modern skepticism and the paranormal are partly defined in terms of science. Therefore, if we are to understand what the paranormal is, or what modern skepticism is, we have to understand the nature of science.

Philosophers and others have been trying to explain science for centuries, mostly without success. We will avoid the problem. Instead of trying to define science, we will do something less ambitious. We will outline a few general processes common to all sciences, and construct a simple model of how science works. This avoids complexities, but tells us something deeply important about science, and also about both skepticism and the paranormal. Incidentally, the approach adopted here might seem new: instead of starting with abstract philosophical principles, it begins with what scientists actually do, which is far more comprehensible.

We will do more than look at the intellectual structure of science. Although science is a very successful enterprise, it has some very real problems. We will look at some of the challenges that science is facing. Although the problems do not threaten the existence of science, they do imply that its spectacular development may slow in the future. They also suggest weaknesses in the structure of science, and it is in some of these gaps that the strange blooms of the paranormal can flourish.

Chapter two moves toward the heart of what the book is about: it looks at the paranormal. We will find that definitions of the paranormal have a strange consequence. They imply that not only is the paranormal currently not understood, but that it cannot possibly be understood at any time in the future. The paranormal must remain forever unknown.

This is an unusual finding, as it means that the paranormal will forever remain on the fringes of intellectual life. It will never be completely accepted, but it can never be rejected either. Almost as an incidental issue, we will also see that the paranormal is extremely diverse. The different types of claim that make up the paranormal are quite different from each other, and in some cases are contradictory.

Chapter three begins to look at skepticism, the key concept in this book. It is clear that modern skepticism is one of the intellectual descendants of a tradition that can be traced back at least 2500 years, to the ancient Greeks. Then we move on to more modern thinkers.

René Descartes is often regarded as the first modern philosopher. As far as modern skepticism is concerned, he is an intriguing mixture of the old and the new. His willingness to doubt everything, to pursue his thought wherever it might lead, is disconcertingly modern and also very brave. His rapid return to the arms of orthodox religious belief marks him as still being part of the age of faith.

Once we come to David Hume, a century later, we are in the modern world. Hume relished doubt and skepticism, and pursued his thoughts as far as he could. His essay on miracles is still a masterpiece and has within it the seeds of the modern skeptical movement.

Chapter four looks at modern skepticism. Although Harry Houdini and Bertrand Russell did important skeptical work in the early twentieth century, it is not until the 1970s that the modern movement was born. Writers such as Martin Gardner and activists Paul Kurtz and James Randi launched the modern skeptical movement, with its focus on the investigation, and, where necessary, debunking of the paranormal in all its forms. The worldwide growth of the skeptical movement has been spectacular, and so has its prominence in the public eye. Here we also look at some key characteristics of modern skeptical thought.

Chapter five is important. The broad sweep of skeptical thought is easy to grasp, but it does not always apply to particular cases. Therefore, we need intermediate ideas which link main skeptical concepts and paranormal claims. Here we do that. We look at important skeptical ideas and see see how skeptics use them. We explain key ideas such as the placebo effect, the double-blind controlled trial and the amazing effects of statistical coincidence.

At this point, the reader will have a good grasp of skeptical thought and how it can be applied. However, more is needed. Why should one use skeptical thought at all? Chapter six addresses this question, raising the disconcerting issue of whether it is unethical to hold certain types of belief, and why this might be so. This field – the ethics of belief – has been dormant for a long time, and is re-emerging into prominence. There is a good case for saying that skepticism is in fact an ethical movement, and here we explain why.

Chapter seven goes beyond the type of skepticism covered in the book so far. It asks whether skepticism has any value in areas beyond the paranormal. Clearly, if the definitions outlined in Chapter four are used strictly, then the answer is no. On the other hand, if we look for areas of knowledge where a skeptical type of approach might be useful, then the answer is an emphatic yes. An obvious area where a skeptical type of approach can be used is in some areas of history. We will look at holocaust denial and see how the skeptical tools resemble those to be found in the paranormal area and how far they differ. We will also cover other areas where skeptical approaches appear to be used, such as conspiracy theories, and it is clear that the skeptical approach has validity beyond the paranormal.

Some people, when acquainted with the skeptical approach, denounce it as being conservative or always in favour of the status quo. In a sense this is true, although it only applies where the status quo has a great deal of evidence to recommend it, and the alternatives little or none. In some cases, however, it is the skeptical approaches that are causing change, and we see one of them in action in the section on evidence-based medicine in Chapter seven. The evidence-based medicine movement insists on using the highest quality of evidence in making decisions about treatments. This is completely different from the traditional approach to medicine, which stressed the primacy of clinical judgement and the right of the professional to treat as he (it was usually he) thought fit. As we will see in this chapter, this implies that skepticism is not a stand-alone movement. It is part of a larger evidence-based movement that is already radically changing some fields (such as medicine), and promises to change many more.

It should be clear by now why this book has something new and distinctive to say. It seeks to clarify exactly what modern skepticism is, what its ethical underpinnings are, and how it translates into practice. This is a large set of goals for one book, and only the reader can decide whether I have succeeded or not. In my view, it is certainly something worth attempting!

1 | The nature of science

YOU MAY FEEL a little cheated by the subject of this chapter. Isn't this book about investigating the paranormal? Why should we spend a whole chapter looking at science? There are at least two compelling answers. First, both the paranormal and skepticism are defined in terms of science: we will see this in Chapters two and three. Without some understanding of science, the other key terms will make no sense. Second, many paranormal practitioners argue that they are in fact doing science. Others are fiercely opposed to the idea. Science is simply too important to be ignored, and we need to know how it works.

What is science? For centuries, philosophers have tried to work out exactly how scientific knowledge differs from other types of knowledge. In general, the attempts have failed. Alan Chalmers (1988) has written a good summary of the major attempts. In this chapter we will outline a couple of simple models of science. In turn, these will enable us to understand some key points about both skepticism and the paranormal.

We might start with the views of one of the greatest scientists of all time. Albert Einstein summarised the entire goal of science in a single sentence: 'The grand aim of all science is to cover the greatest number of empirical facts by logical deduction from the smallest number of hypotheses or axioms.' (Calaprice 1996, p. 178).

Let us unpick this concise statement a bit more, as it tells us a great deal. Einstein makes clear that he is referring to all of science, not just physics, his own area of work. He tells us that there are 'facts,' by which he clearly means well-established pieces of knowledge about the universe, and that these are to be explained by a small number of 'hypotheses or axioms'. Hypotheses or axioms, or theories, to use another word, are explanations concocted by human beings to explain the universe, so Einstein is telling us

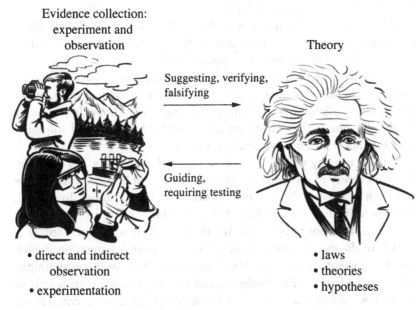

Evidence collection:
experiment and
observation

Theory

Suggesting, verifying,
falsifying

Guiding,
requiring testing

- direct and indirect
  observation
- experimentation

- laws
- theories
- hypotheses

Figure 1.1 Relationship between theory and evidence in the process of science

that science aims to explain as much as possible of the universe by human theories, and that there should be as few theories as possible.

In sum, science consists of at least two human activities. One is collecting evidence – by experiment or observation – and the other is theorising, trying to work out how the world, or some part of it, actually works. We can represent these two scientific activities in Figure 1.1. Einstein, the supreme theorist, stands for the theorising activity in science, and a group of scientists and instruments represents the collection of evidence. We should unpack the meaning of theory and evidence collection before moving on.

In science, statements aimed at explaining how some part of the universe works are called theories. The best-established theories are usually referred to as 'laws'. Isaac Newton produced several of them, including three laws of motion, a law of gravitation and a law of cooling (Koyre 1965). Robert Hooke originated the shortest law known: in a body that is being stretched, tension is proportional to extension (Stone 2003).

Almost as well established as laws are theories. Among the general public, the word 'theory' is sometimes used to mean nothing more than someone's ideas, but scientific theories are not like that. In general, they are large-scale organising ideas, usually with a great deal of evidence to back them. Charles

Darwin's theory of evolution is one example (Oldroyd 1980). Another is Alfred Wegener's theory of continental drift (Le Grand 1990). Both explain important aspects of the world, and in both cases scientists are satisfied that there is enough evidence to support them.

Least general of all are hypotheses. These are specific statements which can usually be directly tested. If I believe that adding two chemicals together will produce sulphuric acid, I am proposing a hypothesis. It is directly testable because if the two chemicals are added together, and sulphuric acid does not result, my hypothesis is wrong.

Usually a theory will yield a whole range of hypotheses. For example, Darwin's theory of evolution yields a whole range of hypotheses about the evolution of whales, the eye, birds' feathers and much more. In general, if a theory yields successful hypotheses, it is likely to be accepted. If the hypotheses turn out to be wrong, the theory is likely to be discarded.

Where do theories come from? This is a complex question, which could lead us into psychology. However, part of the answer is very simple. Scientists look at existing evidence, and try to work out what it tells us about how the universe works. There is nothing unusual or eccentric about this process, simply clever people trying to understand what is going on.

On the other side of Figure 1.1 is the collection of evidence. Usually, scientists like to collect evidence through carefully designed experiments. These yield the most reliable evidence, and the most useful. However, some sciences are simply not amenable to experiments. Cosmologists, for example, cannot experiment with the formation of galaxies. For a range of sciences, experiments are usually impossible, and careful observation and examination must serve as a substitute.

We now have the two major components of science clear in our minds. Now the question which goes right to the heart of science: What is the relationship between theory and evidence in science? This is the problem on which philosophers and historians have spent endless effort and thought. Francis Bacon (1856 [1620]) thought that observations accumulated and led to the formulation of more and more general statements through a process which he termed 'induction'. Karl Popper (1968), on the other hand, argued that theories should be stated in ways that made them capable of being falsified so that evidence could decide whether they stand or fall. Chalmers (1988) describes these ideas, and a good many more, but we will be less ambitious. Our goal is simply to make some basic statements about the relationship between theory and evidence in science.

First, in general, the collection of evidence is guided by theory. It is impossible – and pointless – to collect evidence without some sort of theory in mind. Scientists are human, and a scientist who has just formulated a new theory may well begin to collect evidence in support of it. Equally, scientists opposing a theory may conduct experiments that they believe will prove it false. Whatever the motive, the collection of evidence is guided by theory.

A famous example of this guidance is the way in which the noted physicist Arthur Eddington collected evidence on the general theory of relativity (Singh 2004). Einstein had worked on his general theory for ten years, from 1905 to 1915. When he published his theory there were very few ways it could be tested. However, the theory did predict that the light of stars which passed near to the Sun would be 'bent' by the Sun's gravity, and this could be measured during total eclipses.

Eddington and a party of scientists headed for Africa, where a total eclipse would be visible from the depths of the jungle. After many adventures they took the relevant photographs, and returned them to Britain to be analysed. Einstein's theory guided Eddington and his colleagues to a remote location, where they photographed the positions of stars near the Sun during the eclipse. Careful study of the photographs showed that Einstein's theory was verified: the stars did indeed seem to be out of position, and by roughly the amount predicted. Einstein's theory stood – and still stands.

As this example shows, if a theory leads to valuable predictions and hypotheses, scientists will begin to regard it as established. If it keeps failing in these respects, it will eventually be discarded. In the case of Eddington, the stars' light did appear to be bent by the Sun's gravity, and by the amount Einstein's theory predicted. This verified Einstein's theory. Hence, there is a strange, circular relationship in science between theory and evidence. One guides the collection of the other: one determines the shape of the other. And, gradually, through this mechanism, the nature of the universe is made clear.

Who makes these decisions? Who decides whether a theory has been supported by evidence or falsified? There is no official body to decide these matters. Instead, there are communities of scientists who work on the problems, study the different theories and arrive at conclusions about the meaning of the evidence. This has been a feature of science since the seventeenth century (Burns 1981).

ADDITIONS TO THE MODEL

We could elaborate the simple model of science in many ways. However, it is useful to make one general observation and to add one more feature. The observation is this. We can see from the model that theory in science is based on evidence. Evidence decides whether theories stand or fall. It follows logically that when scientists are formulating theories, they want the best evidence available. It would be an appalling waste of time to formulate theories based on evidence that is not reliable. Therefore, scientists usually insist that the evidence they accept must be verifiable or, even better, replicable. The latter type of evidence is the best of all. It denotes experiments or observations that can be performed by anyone and that will give the same result. Earlier in this chapter, we talked about my hypothesis that mixing two chemicals would produce sulphuric acid. To be generally accepted, the experiment should be replicable. Anybody should be able to mix those chemicals and produce that result. Of course, some sciences do not permit experiments. In this case, the results should at least be verifiable: they should be checkable by other people. For example, once Einstein's theory had made the prediction about the Sun bending the light of stars, anyone could have done the observations – though only Eddington and his colleagues actually did. It is for this reason that scientists are insistent upon the highest possible quality and reliability of evidence. To construct a theory on poor evidence is as futile as building a house on sand. In both cases the edifice is doomed and much time will be wasted. As we shall see, this has happened in some areas of science, and the results have been catastrophic.

The single extra feature is the effect of technology on science. We are used to thinking of science as influencing technology, and so it does. However, technology affects science in several ways. The most important of these is through the collection of evidence. Advancing technology enables the collection of evidence, often in areas where nothing was known before. Radio telescopes in the 1950s and 1960s completely transformed our view of the universe. They could see much further than optical telescopes, and could 'see' electromagnetic wavelengths inaccessible to the eye. Another example, in the same time period, was the development of ocean floor geological drilling. This technique, for the first time, enabled the geology of the seabed to be understood, and precipitated the scientific revolution of plate tectonics (Le Grand 1990). More recently, space probes, giant atom smashers and DNA analysis techniques have also enabled new types of scientific evidence to be collected. Indeed, at least one major historian of science has suggested that scientific progress is intimately bound up with technological changes, such as instrumentation (Price 1984).

This modification results in the simple model outlined in Figure 1.2. Science guides the selection of evidence which, in turn, dictates which

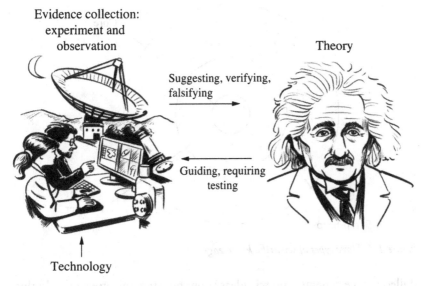

Evidence collection:
experiment and
observation

Theory

Suggesting, verifying,
falsifying

Guiding, requiring
testing

Technology

Figure 1.2 Addition of technology to the scientific model

theories stand and which are discarded. At the same time, developments in technology enable evidence to be collected over wider and wider areas, thus giving theorists more and different evidence to work with. Bigger telescopes, and different kinds, such as radio telescopes and orbital telescopes, mean that new evidence about the universe is available to astronomers. New and better laboratory apparatus means that chemists and biologists can perform better experiments. We can therefore regard the development of science as a kind of widening spiral, moving ever outwards into new areas, and, usually, leading to the understanding of more and more phenomena. At the same time, the precision and reliability of explanations in the areas covered by science constantly improves.

A WAY OF LOOKING AT SCIENTIFIC KNOWLEDGE

The simple model we have developed is useful in understanding some of the basic dynamics of science. It highlights the fact that science is not a stock of established facts, nor is it some abstract process, mechanically generating knowledge. Instead, it is a very human process by which highly

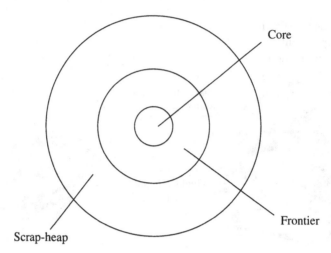

Figure 1.3 Three types of scientific knowledge

skilled people generate and test ideas about how the universe works. In this section, we will develop another simple model concerning science; this one looks at the scientific knowledge that is produced.

Some years ago, the distinguished sociologist Stephen Cole (1992) suggested that scientific knowledge could usefully be divided into 'core' and 'frontier' types. In 1998 this author (Bridgstock 1998b) added to this the concept of the scientific 'scrap-heap', and these three terms give a good picture of scientific knowledge at any moment.

Figure 1.3 summarises scientific knowledge in a very simple fashion. At the centre of science are the theories and findings of the 'core'. This is knowledge which the scientific community regards as established. Examples would include the theory of evolution, plate tectonics, the Big Bang origin of the universe and quantum theory. In each case, the knowledge is sufficiently established that it can be taught in schools and undergraduate university courses. In principle these theories can be overthrown by contrary evidence, but scientists regard the current state of knowledge as justifying a high level of confidence in their correctness.

Outside the core is the 'frontier'. Here are theories which have not yet been fully justified by evidence, or which remain controversial. There might also be unverified experiments and observations. Some of the claims in this frontier area will eventually be established and become part of the core. Others will be relegated from serious consideration, and become irrelevant to scientific investigation. Because of their uncertain future, these will not normally be taught, except perhaps in advanced courses at university.

Almost all theories, when first proposed, appear in the frontier area of science. Some are rapidly accepted into the core, others, such as continental drift, may spend decades in this uncertain state. It depends on the evidence collected, and the arguments which take place within the scientific community.

Finally, outside the realm of scientific knowledge is the 'scrap-heap'. This is the area of theories and research findings that have been rejected by the scientific community. It was once believed that the Earth stood at the centre of the universe: now that theory is taken seriously by almost no one. It was once believed that the continents did not move around. Now that theory, fixism, as it used to be called, is on the scrap-heap (Le Grand 1990). The colour theory of quark interactions (Pickering 1980) and Lamarck's theory that acquired characteristics can be inherited (Gardner 1957, pp. 140–2) are also on the scrap-heap. We should note also that some experimental results are here too. Relativity theory is based on the assumption that, wherever you are and whatever speed you travel at, the speed of light always appears constant. However in the 1930s a reputable scientist named Miller conducted a series of experiments which appeared to show that this was incorrect (Gardner 1957). These results are now discounted: nobody knows quite what went wrong. So are the notorious 'cold fusion' experiments, which purported to show that elements could be fused easily, and without a great release of energy (Close 1992). In principle, new discoveries could bring theories and even experiments back from this graveyard of science. In fact, few ever return.

This view of scientific knowledge gives us a snapshot of the state of science at any moment. Over time, new theories are proposed and make their way into the core, or are relegated to the scrap-heap. Over time, the core grows ever larger as science presses further into the unknown, increasing understanding about the nature of the universe. So, probably, does the scrap-heap, as more and more theories are proposed, tested, and shown to be wrong.

SOME LIMITATIONS OF THIS VIEW OF SCIENCE

The model will prove useful, but it has a number of flaws, and we might note a couple of them now. First, some scientific theories are not relegated to the scrap-heap, but are subsumed under more general theories. This happened to Newtonian physics. In the twentieth century, it has become clear that Newtonian predictions simply do not work for objects moving with high velocity (Singh 2004). Newtonian physics still exists, but it is treated as a special case of relativistic physics. NASA still launches and manoeuvres its spaceships using Newtonian equations, but scientists know that they have strictly limited validity.

Second, this model only works for pure science. This is science carried out for the purposes of understanding the universe better (Bridgstock 1998a). It does not relate to applied science, where the goal is to do something practical. Applied science is often done in business corporations, to develop new products or to eliminate problems in existing ones. It follows that the priorities will be quite different from those outlined here.

A third limitation of this model is that it presents science as though theories are largely independent of each other. Of course, there are thousands of theories in science, and there is evidence bearing on all of them. Sometimes theories in one field have implications for another, and controversies can erupt across disciplinary lines.

A famous example of this occurred towards the end of the nineteenth century. Geologists and biologists – most notably Charles Lyell and Charles Darwin – had formulated theories about the formation of the Earth and living things. They had buttressed their theories with great amounts of evidence. However, the distinguished physicist Sir William Thomson (later Lord Kelvin) had done physical calculations on the age of the Earth, using cooling theories. He had found that the Earth was only some twenty million years old. This caused great problems for geology and biology, as the processes in these fields suggested that the world had to be hundreds of millions of years old. The controversy rumbled on for years, until the great physicist Rutherford established that radioactive materials could delay the cooling of the Earth, giving the other scientists the time their theories needed (Gorst 2001). Thus, theories in science are not isolated. They exist in a network of other theories and evidence, which overall must be consistent, and which give a picture of the universe and the way it works.

Historian of science Thomas S Kuhn (1966) gave a dramatic account of the rise and fall of large-scale theories in his groundbreaking work *The Structure of Scientific Revolutions*. Here, Kuhn outlined the concept of a 'paradigm' – a large perspective within which many theories can be accommodated. For much of the time, scientists work within these paradigms, solving scientific problems and clarifying the implications of theories. However, occasionally a paradigm may be beset by many anomalies – results and findings that cannot be explained – and this can lead to the eventual destruction of the paradigm, and its replacement with another. These are the 'scientific revolutions' in Kuhn's titles, and examples include the rise of the heliocentric model of the solar system (Kuhn 1957) and the gradual acceptance of continental drift (Bridgstock 1998b, pp. 88–94)

THE HUMAN SIDE OF SCIENCE

What does all this tell us about the nature of scientific knowledge? First, the advance of science is in fact a very human process. Large numbers of (usually) very intelligent people are struggling to understand the nature of the universe. If they stopped trying, the advance of science would grind to a halt. It follows that we would expect all the usual characteristics of human behaviour within science: brilliance and pig-headedness, selflessness and egotism, inspiration and all the other characteristics that help make us human. Science is a human activity, and it bears all the characteristics of humanity.

Second, to the outsider, scientific knowledge looks strange. On the one hand it is highly reliable, on the other hand it is always provisional, and can always be falsified by new evidence. However, this paradox is perfectly logical. It is *precisely because* scientific knowledge can always be falsified that it is so reliable. Scientific laws and theories which have survived years of testing and checking are bound, by that very fact, to be well verified and likely to be supported by any further research.

There is another important point which follows from the basic nature of science. One often hears of science 'proving' something to be so. In fact, the very nature of science makes this impossible. Nothing can ever be conclusively proved by science, since the final product is always provisional. Compare this to, say, Pythagoras's Theorem. Granted certain assumptions (known as axioms), one can prove Pythagoras's Theorem to be true. However, Newton's Laws cannot be proved in the same way: they are accepted because they (usually) fit the evidence, and can always be discarded if they do not.

This makes scientific knowledge radically different from other forms of knowledge, even though it shares some attributes with them. Few other forms of knowledge are at the same time highly reliable and completely provisional. None are as prestigious as science, and yet at the same time provide explicitly for their own destruction. Science is a unique form of human knowledge, even though it partakes completely of its human origins.

It should also be clear from this model why the faking of scientific results is such a grave crime. Scientific theories rely on evidence for their formulations. If the evidence is unreliable, or, worse, simply false, then the theories are bound to be false, and there is great damage to the scientific enterprise. A good example of this damage is the conduct of the Indian geologist VJ Gupta (Broad and Wade 1985). Over many years, Gupta built up an awesome reputation as an expert on the geology of the Himalayas. He would approach well-known scientists with geological specimens and accounts of where in the Himalayas he had found them. Papers would be written jointly by the two scientists. In due course, Gupta built up a bibliography of over 300 books and papers. However, Gupta's evidence was faked. In reality he had bought specimens in Africa and China, and his stories about the Himalayas were false. As a result, scientific knowledge of that part of the world has been terribly damaged. It is a classic example of how science needs reliable evidence to function properly.

IS SCIENCE ALWAYS RIGHT?

It is fairly easy to find cases where the scientific process does not seem to have led to the best answers, at least in the short run. For example, in 1915 Alfred Wegener put forward his theory of continental drift – the idea that the continents move about on the Earth's surface. It was not until about 50 years later, in the mid 1960s, that Wegener's ideas were accepted (Bridgstock 1998b). Wegener died before his theory was established, and many other scientists have gone without recognition for years. For example, for many years scientists were trying to explain exactly how plants absorbed sunlight and used it to build the necessities of life for themselves. Most people thought that some sort of chemical explanation was correct, and indeed this is the most obvious view to take. However, Peter Mitchell, a British scientist, argued that a physical explanation was correct. For about twenty years Mitchell's ideas were ignored, and behind his back he was denigrated. However, further research showed that Mitchell was correct, and eventually he was awarded the Nobel Prize in recognition of his work (Magill 1990).

The case of Barbara McClintock is similar in many ways. For decades McClintock was heavily immersed in plant genetics, and she eventually formed the view that genes are not fixed on DNA, but can move around. This was ignored for many years before being finally accepted, and McClintock was well into her eighties when she received the Nobel Prize (Keller 1983).

We could look at other cases, but the three examples mentioned here illustrate the point. Do they show that science can often be wrong and unjust? In all cases, people and ideas deserving of recognition were spurned for long periods of time. Science is a human process, and some human ideas, at any given time, are likely to be wrong.

At the same time, these three cases do not condemn science at all. In the case of Wegener's theory, there were profound problems with it for many years. Perhaps the most important problem was that there was no convincing mechanism by which continents could move around. A continent is a large object, and it would take an enormous force to move it. Further, continents are made of relatively light, brittle material compared to the underlying rocks of the Earth's crust. Therefore, any force large enough to push the continents through the crust would also be strong enough to tear the continents apart.

This is a strong objection. Although Wegener and his supporters produced a great deal of evidence that the continents had moved, it seems perfectly reasonable for scientists to refuse to accept the theory until some adequate mechanism had been shown to exist. This mechanism was discovered in the 1960s, with the advent of knowledge gained from deep-ocean drilling. This new knowledge showed that the surface of the Earth is divided into large plates which are in constant motion. They carry the continents, and are in constant gradual motion, sometimes colliding with each other, and sometimes sliding one over another. What keeps these plates in motion are enormous convection currents in the Earth's crust. Thus, a force exists which is more than strong enough to carry continents, and because it works on the geological plates, not on the actual continents, it does not destroy them.

Before the 1960s it was perfectly reasonable for scientists to be doubtful about the claim that the continents moved. After all, in the absence of a motive force, or of an explanation as to how the continents would not be destroyed, skepticism was a logical stance. After the 1960s, however, opinion shifted decisively in favour of continental drift – or plate tectonics, as it is now known. This episode, therefore, does not reflect badly on science at all. Scientists had good reasons for not accepting continental drift, but once the problems were resolved, the resistance melted away (Le Grand 1990).

But what about the two scientists, Peter Mitchell and Barbara McClintock, who were denied rightful recognition for many years? Does this not show that science can be unfair and, sometimes, just plain wrong. The answer is both yes and no. Both of these scientists went unrecognised for long periods. In addition, there are embarrassing reasons as to why this may have been so. Peter Mitchell had given up an academic position and worked in a small independent research centre. He was an outsider to mainstream science. Barbara McClintock was a woman, and for much of her career, women were simply not accepted as being capable of doing excellent science. It is worth pointing out that both Mitchell's and McClintock's work was eventually accepted, and they were honoured for their excellence. Their work was not permanently consigned to the 'scrap-heap', its acceptance into the core of science was delayed.

This raises the question of whether science can ever permanently delay the acceptance of good work. That is: can worthwhile theories or evidence be excluded permanently from science? In principle it could happen, although it is difficult to show that any good scientific work has ever been permanently rejected. We should be wary of claiming that science is always just, logical and right, but we should require very strong evidence before concluding that it is largely wrong.

Most scientists have developed defences against people who claim to be 'misunderstood geniuses'. By one means or another they contrive to exclude or ignore these people, and to get on with their own work. At the same time, as we have seen, the valuable work of outsiders like Mitchell, McClintock and Wegener was eventually recognised. One can also point to the acceptance of the Indian mathematical genius Srinivasa Ramanujan. (Kanigel 1991). Here was a complete outsider to western mathematics, a young Indian clerk, who was accepted into Trinity College, Cambridge, to work with some of the best mathematicians of the day. His genius, after some hesitation, was fully recognised. Although scientists may not always be right, it is hard to sustain a charge of systematic discrimination against unorthodox ideas. In general, science appears to be genuinely what it claims: an enterprise that seeks to explain the workings of the universe, and seeks to make all of its explanations testable. The position of the skeptics, and of this book, is that this is usually the way science is. Scientists do not arbitrarily rule out explanations, nor ignore worthwhile theories or important findings, except in odd cases. If someone claims that this has happened, the onus is on them to establish it.

SCIENTIFIC KNOWLEDGE IS HELD BY A COMMUNITY

One other important point follows from the picture of science we have spelt out above. Scientific knowledge enters the 'core' when it is accepted as being established by the scientific community. No one person's beliefs – no matter how strongly held – can establish knowledge in this way. Thus, it is part of the scientific process for scientists to be able to persuade other practitioners of science that their theories are correct, or that their observations are accurate.

Naturally, this need for argument and evidence means that scientists can be subjected to extremely tough, searching criticism. Carl Sagan (1997, pp. 34–5) describes some of the features of routine, ongoing criticism among scientists: 'You sit in at contentious scientific meetings. You find university colloquia in which the speaker has hardly gotten thirty seconds into the talk before there are devastating questions and comments from the audience.' Sagan goes on to describe the rigours of the peer review process for scientific articles, and concludes: 'Instead, the hard but just rule is that if the ideas don't work, you must throw them away.' Overall, Sagan believes, 'Valid criticism does you a favour.' (Sagan 1997, p. 35)

This is a tough way to proceed, and it can be distressing. However, as Sagan says, it is necessary. To be strongly established, a theory must be supported with evidence and, as we have seen, that evidence needs to be reproducible. That is, if one scientist claims that a certain experiment leads to a certain outcome, other scientists must be able to perform the same experiment and achieve similar results.

What if they cannot? What if some scientists verify an experiment's results, and others do not? Usually, the result is a kind of limbo, with different people taking different positions. This happened when Joseph Webber claimed to have detected gravity waves with equipment that was generally thought to be too insensitive (Collins 1975, 1981). It also happened when Fleischmann and Pons claimed to have produced 'cold fusion' in the laboratory (Close 1992). In both cases some other scientists claimed to have replicated the results, and others did not. The result in both cases was a confused argument, with assorted features of the different experiments being pointed to as reasons for the strange results. In both cases, incidentally, the claims were finally rejected.

It does seem as if one of the strongest recommendations for the truth of a particular experimental outcome is if another scientist, either opposed to the claimant, or at least neutral, can achieve similar outcomes with replication. This certainly happened with the famous Michelson–Morley experiment to determine if the speed of light varied (Singh 2004). The astonishing result – that the speed of light does not vary regardless of how fast, or in what direction the experiment is moving – was replicated repeatedly, and was finally accepted as being correct (Gardner 1957).

To leap ahead a little, we might consider possible outcomes if a group of researchers claim certain important results in their field. Other researchers who do not accept the group's views replicate the experiment and find no significant result. After a prolonged dispute, the first group makes the claim that the very presence of unbelieving people renders the experiment impossible. That is, only devout believers in the experiment's results can produce positive outcomes. From our discussion of the nature of science, it seems clear that the researchers making the claim are doomed to non-acceptance. The best kind of evidence is supporting evidence from hostile or neutral people, and these researchers cannot provide this. As we will see, this is a key claim of parapsychologists, and is a partial explanation of why their field has never achieved full scientific recognition.

There is another way in which that model tells us something about the paranormal and science. Note that in order to work, science needs to have both the theory and the evidence in working order. Relevant and reliable evidence needs to be flowing in, and scientists need to be testing the theories and, if necessary, putting forward new theories to explain the evidence.

PROTOSCIENCE AND PSEUDOSCIENCE

Clearly, if either component of science is not working – theory and evidence collection – then science cannot work. If no new evidence is coming in, then there can be no testing of new theories, and so useful new theories cannot appear. It is likely that scientific thinking will degenerate to questions of the 'How many angels can stand on a pin?' type, rather than creatively seeking to expand understanding of the universe. Thus, science needs new evidence, or it cannot function. It is equally important to have new theories that can grasp, and make sense of, the new evidence. Indeed, if theory cannot make sense of evidence, then research is in real trouble. Imagine that new evidence is produced in some field of research. However, although the evidence is reliable, it makes no sense at all in terms of theory. That is, it seems impossible to explain what the evidence means. In this case, again, science grinds to a halt. Since no sense can be made of the results, it is impossible to do further research. One can collect more evidence, but it is impossible to tell whether it is increasing scientific understanding, or whether it simply adds more numbers to an incomprehensible heap. This is what seems to be happening in the field of parapsychology, as we shall see. Research is being done, but there seems to be little or no advance in understanding.

What happens to a claim – paranormal or otherwise – which has evidence but no theory? It would seem likely that it goes into a kind of grey zone. Assuming that the evidence is reliable, it cannot be integrated into the body of scientific knowledge, and it cannot even really be regarded as part of the frontier. It remains an uncomfortable anomaly, not rejected, but not really suitable even as a field for research.

Sometimes a science does not appear to be working properly for historical reasons. Very early in the development of a new area the theory may fit the evidence poorly, or the evidence may be of very poor quality. It seems clear that there is something worth studying, but nobody is very sure what. We can call this sort of research 'protoscience', meaning a science in its very early stages. We might regard the study of plate tectonics between about 1917 and 1960 as being in this state: whether the continents moved or not was under discussion, and the meaning of the evidence unclear (Le Grand 1990). In hindsight, it is easy to pick out emerging areas of science: theories are put forward and tested and, over time, progress can be discerned.

At the time it is much less clear, especially when there are many imitation sciences that can never become fully scientific. These can be termed 'pseudosciences'. Martin Gardner (1957) essentially started the modern skeptical movement with a scathing review of pseudoscience. The essential feature of pseudosciences is that they lack any of the characteristics of proper science – especially the willingness to discard useless theories – but still claim the status of science. An example might help, especially since in my view it is the most dangerous pseudoscience currently in existence.

AN IMITATION SCIENCE AND HOW IT DIFFERS FROM THE REAL THING

At this stage, we might mobilise a little of our knowledge, and look at a case of an imitation science. Quite explicitly, creation science was set up as an alternative to the scientific view of the universe (Numbers 1992). Its supporters argued that it was as scientific as orthodox science and that therefore it should be taught in schools alongside ordinary science.

Why would anyone do this? To understand the origins of creation science, we have to understand modern Christian fundamentalism. In many countries, a considerable minority of the population believes that the Bible is not only a holy book, but is the precise and infallible word of God. If one looks at the first book of the Bible, Genesis, one finds plain statements about the origins of the world, the universe and humanity: God created them all in six days. Therefore, the Christian fundamentalist is faced with a difficult choice: accept that the Bible is not totally, literally, true, or reject large parts of the findings of modern science. Some fundamentalists have chosen the former course (Numbers 1992), and some have chosen the latter.

There are a great many problems with choosing to believe the Bible literally. Distinguished theologian James Barr has produced a whole book of them (Barr 1984). Somehow, fundamentalists who believe the Bible to be literally true also have to reconcile themselves to living in a world that accepts quite different views. In particular, the Darwinian theory of evolution causes fundamentalists much distress. They believe, as the Bible tells them, that they are a unique creation, made in God's image. This is irreconcilable with the millions of years of gradual development which orthodox science tells us have led to our current state of development. Further, science paints a picture of a universe billions of years old, which began with a primeval cosmic fireball, and exploded into its current form. Again, this clashes with a literal reading of the book of Genesis.

The large fundamentalist community in the United States tried a number of measures to cope with science. For a while, the teaching of evolution was banned in some American states. Then attempts were made to teach religious views alongside scientific ones (Numbers 1992). For various reasons, none of these were successful. So, beginning in the 1960s, fundamentalists began to use the strategy of demanding 'equal time'. They argued that the Biblical account of creation offered an account of origins just as scientific, and just as convincing, as the orthodox scientific account. Instead of an ancient Earth, they proposed one only ten thousand or so years old. Instead of billions of years of evolution, they proposed six-day creation. The geological strata of the Earth, they argued, were laid down by Noah's Flood, and humanity was dispersed into many races and tongues during the Tower of Babel incident.

To support their view, creationists wrote textbooks, such as Henry Morris's *Scientific Creationism* (Morris 1984). They founded 'scientific' research organisations, produced course materials (e.g. Bliss 1976) and pumped out a mass of promotional materials, magazines and newsletters. The goal was not to convince scientists that their views were wrong. The targets were legislators and the public. The aim was to pass laws ensuring that creation, since it was an equally good scientific theory, should receive equal teaching time in schools.

For a while during the 1980s, the creation scientists appeared on the verge of success. At one stage 23 American states were considering legislation mandating 'equal time for creation' and in two states, Arkansas and Louisiana, these laws were actually passed.

We will return to creation science later in the book. The aim here is to look at creation science through the simple models outlined above. On the basis of what we know about creation science, can it be science in the same sense as the science talked about above?

Commentators and critics of creation science began to research the creation movement. They were disconcerted by what they found. Usually, creationist organisations fixed the beliefs of their members by prior decree. For example, I was concerned at the growth of influence of the Creation Science Foundation in the Australian state of Queensland. A search of the Articles of Association of the Foundation produced a series of statements which members were expected to hold. Among them were the following:

> The scientific aspects of creation are important, but are secondary in importance to the proclamation of the Gospel of Jesus Christ, the Sovereign Creator of the Universe and Redeemer of Mankind.

and

> The Bible is the written Word of God. It is inspired and inerrant throughout and the supreme authority in all matters of faith and conduct. Its assertions are historically and scientifically true in all the original autographs.
>   (Bridgstock 1986b, p. 81)

From which it follows, of course, that the account of origins presented in Genesis must be literally true. Apparently to prevent any questions about what the Bible and Genesis say, the Articles of Association also go into considerable detail about exactly what members are expected to believe. It is stipulated, for example, that the days in Genesis are 'six (6) consecutive twenty-four (24) hour days of Creation' (Bridgstock 1986b, p. 81). It is also stated that Scripture teaches a recent origin for man for the whole Creation, that Noah's Flood laid down much 'fossiliferous sediment' and that death entered this world as a consequence of man's sin.

We now know enough about creation science to be able to talk about whether it can be regarded as scientific or not. The first point is a simple one. The goal of science is to understand the natural universe. However, the goals of creation science are clearly the proclamation of a set of religious beliefs. Science, in the view of creationists, is secondary to religious belief.

The second point concerns the listing of beliefs which it is necessary for creationists to hold. In the creation model of science, the theory is effectively fixed: the Bible is regarded as a 'reliable framework for scientific research'. In the simple model presented in Figure 1.1, we saw that theory can be overthrown by contrary evidence. In the creation model it cannot: the framework of theory is fixed, regardless of the evidence.

For these two reasons alone – the different aims of creation science and the unalterable status of its theory – we are justified in saying that creation science is not science. It is something quite different. This is exactly what was concluded by Judge William Overton in an Arkansas court case concerning creation science. Judge Overton discussed the nature of science, in terms similar to (but more complex than) those outlined here, and concluded that creation science was not scientific at all (Overton 1984). For this reason, because of the very basis of creation science, we are justified in concluding that it is not a protoscience but a pseudoscience.

However, there is another point that comes from this view of science. If we look at Figure 1.3, we can ask exactly where does creation science's theory fit? What sort of knowledge is it? The answer is strange, and revealing. First, the idea that the world was created by God, as described in Genesis, was the predominant theory in the sciences of geology and biology until well into the nineteenth century. As more and more evidence accumulated, the theory came under pressure. Finally, the geological theories of Charles Lyell (1830–1833) and the evolutionary theory of Charles Darwin (1859) replaced creationism. From the scientific viewpoint, creationism belongs on the 'scrap-heap'.

As we have seen, a theory is not completely doomed by being relegated to the scrap-heap. It could be resurrected by the accumulation of new evidence indicating that it does have, after all, valuable explanatory power. If this happened, we would expect to see creationism making its way into the frontier of science, as the evidence appeared. Finally, probably after a good deal of controversy, it might enter the core of science.

However, creationism did not follow this pattern. First, very little new evidence was produced by the creationists themselves. Second, they attempted to have their theory promoted from the scientific scrap-heap directly into the scientific core by legislation. That is, decisions about what was reliably known in science were to be made, not by scientists but by politicians responding to political pressure. This is so much at variance with normal science that, on this ground too, we are entitled to regard creationism as non-scientific.

In sum, although the models of science are simple, they enable us to see clearly the differences between creationism and normal science. A look at the basis of creation science reveals that its goals are quite different from the goals of science, and, unlike science, its theories are fixed and unchangeable by reference to evidence. In addition, instead of arguing the merits of their views on the basis of scientific evidence, creationists sought to have them promoted through legislation. Whatever creation science may be, it is not science, and the term pseudoscience seems to fit it well.

THE IMPACT OF SCIENCE

What of the effect of science on the larger society? Broadly speaking, science seems to have two main effects. One is the effect which science has on our view of ourselves and our place in the universe, the other is through its impact upon technology (Bridgstock et al 1998b).

Science constantly affects our views of ourselves and the universe. Often, the two go together. For example, during the scientific revolution, scientists like Copernicus, Galileo and Kepler dethroned the Earth – and humanity – from its place at the centre of the universe. No longer did everything revolve around the Earth: after the revelations of science, the Earth was simply one of the planets revolving around the Sun. The universe became much bigger, and our place in it much smaller.

Another, equally shattering impact on our view of the world and ourselves came with Darwin's theory of evolution. This showed the history of life on this planet as a constant struggle for survival, and it situated us within that struggle. Before Darwin, humans could think of themselves as apart from nature. After Darwin, we were a part of it.

The impact of science upon our thought is all pervasive. It has now spread to all aspects of our culture. For example, Peter Watson, in his monumental 800-page intellectual history of the twentieth century attempts to sum up exactly what the twentieth century was about: 'Our century has been dominated intellectually by a coming to terms with science' (Watson 2000, p. 2). Watson does add that this 'coming to terms with science' was often a coming to terms with what people took to be science, although sometimes they did not understand it. Even so, it is clear that science exerted an enormous influence on all kinds of thought in the last century. There seems no doubt it will be at least as important in the century that is just beginning.

The second way in which science impacts upon the larger society is much more visible. It is through its impact upon technology. Scientific discoveries constantly provide the means for technology to develop, creating new products and processes that constantly transform our lives. The most dramatic of these is also one of the best known. Soon after the Second

World War, two scientists in the Bell Telephone Laboratory discovered the semiconductor effect. Within a few years, the first technological impact was occurring: radios shrank from being large boxes with fragile valves inside to tiny devices that could fit inside a pocket. Then, as semiconductors were developed into integrated circuits, computers began to shrink, from huge sets of cabinets to desk-size machines, and finally again to objects which could fit inside a pocket (Weber 1996). At the same time the power of these machines greatly increased, transforming our lives in ways which were – and are – mostly unforeseeable.

Linked with the increasing power of computers has been the development of communications technology. A couple of centuries ago it took months to travel across the world. Now, we can travel from England to Australia in hours, and can communicate from one country to another within seconds. And it is not only in information technology that the changes are so radical. The pharmaceutical and medical sciences are increasing our life spans, curing and controlling diseases which have been uncontrollable for millennia, and offering the prospect of tailoring organisms to suit our needs.

Staggering as they are, these changes should not be regarded as entirely beneficial. Computers are fascinating and enhance work productivity, but they also create unemployment. Pharmaceuticals can save lives, but may also cause suffering through side-effects and interactions. Genetic modification of organisms may assist agricultural productivity, but may also endanger the environment. Almost nothing is an unmixed blessing.

Science has other facets as well, and has its own problems. The sheer scale of science is hard to comprehend. Around the world, it seems likely that about a trillion (a thousand billion) Australian dollars a year is spend on scientific and related research (R and D, as it is called), and the total scientific workforce appears to be well over three million people (National Science Board 2004). This is an enormous industry, and it is not surprising that its impacts are so dramatic.

The sheer size of science seems to be leading to one of its major problems. To be good at science, a minimum requirement is considerable ability. However, a career in science is not all that attractive to intelligent young people, and there are problems recruiting new personnel. One reason is simply that in order to be able to do science many years of arduous study are required. Carl Sagan provides an example of this, discussing the work required to be able to *begin* studying quantum mechanics. He starts by pointing to a long list of mathematical topics that need to be understood, including 'differential and integral calculus, certain special functions of mathematic physics, matrix algebra and group theory' (Sagan 1997, p. 237). How long is this likely to take? Sagan spells it out:

> For most physics students this might occupy them from, say third grade to early graduate school – roughly fifteen years. Such a course of study does not actually involve learning any quantum mechanics, but merely establishing the mathematical framework required to approach it deeply.
>
> (Sagan 1997, p. 237)

With a course of study as harshly demanding as this, and with the current fiercely competitive economic climate, it is not surprising that many young people are opting not to study science. Given the ability and determination to undertake tough studies of this kind, it would be far more profitable to study medicine, or law, or accountancy or engineering. These studies lead far more directly to employment and, in general, to higher levels of pay.

In view of this, it is not surprising that many of the developed nations cannot find enough young people to fill their scientific research places. In the United States, for example, up to half of the science and engineering positions are taken up by non-American immigrants (Stossel 1999). In common with many developed countries, the United States is drawing on the talent of less developed nations to fulfil its own scientific needs. In the longer run this will probably not work. Large developing nations like China and India are now visibly charting a course towards full economic development. Eventually, they will almost certainly have a full range of science institutions, and industries based upon the work of scientists. In short, the supply of bright young people from poorer countries is likely to dry up.

Eventually, I suspect, we are going to have to recognise that the world has a limited supply of the best scientific talent. Some parts of science will be underpopulated with researchers, while other areas will have all they need. It will be a different scientific future, with knowledge expanding more slowly, and serious problems arising about exactly what areas should be given priority for further research.

Because science is complex and difficult to understand fully, it follows that large numbers of people in the general population do not understand it at all. Polls show that the level of knowledge of simple scientific propositions is quite appalling. In the United States, only 17 per cent of the population consider themselves to be well informed about science. And, for each of the statements below, less than half of the population was aware that they are true:

The earliest humans did not live at the same time as dinosaurs.
It takes the Earth one year to go around the Sun.
Electrons are smaller than atoms.
Antibiotics do not kill viruses.
Lasers do not work by focusing sound waves.

(National Science Foundation 2006)

Intellectually, science appears to be in fine shape. New discoveries are constantly being made, and a constant flow of new technologies makes its way into the larger society. However, science is increasingly alienated from the rest of the population. The days of the amateur scientist, able to contribute at the highest level of research, are well and truly in the past.

Another problem is already beginning to make an impact on science, and is likely to impact further. This is the simple fact that science is spelling out a particular view of the universe, and that view is unattractive at best, and horrific at worst, to many people.

We should understand that many scientists do not subscribe to the view spelt out below. They often have strong religious views, or simply do their scientific work without worrying too much about the overall picture. When one looks at the big picture traced out by science, however, it seems bleak. On the individual level, Paul Kurtz outlines the scientific picture as follows:

> But all the works of human beings disappear and are forgotten in short order. In the immediate future the beautiful clothing that we adorn ourselves with, eventually even our cherished children and grandchildren, and all of our possessions will be dissipated. Many of our poems and books, our paintings and statues will be forgotten, buried on some library shelf or in a museum, read or seen by some future scholars curious about the past, and eventually eaten by worms and molds, or perhaps consumed by fire.
>
> (Kurtz 1986, p. 18)

This is sad enough: all our human concerns will, in the long run, prove futile and of little consequence. All of those we know and love will die. However, this is on the individual level. Bertrand Russell takes the same approach on the level of humanity and the effect is even harsher. He begins by pointing out that, from the scientific viewpoint, humanity was produced by causes which had no plan, and that all of humanity's history is essentially the product of accident. As for humanity's future, science tells us:

> that all the labours of the ages, all the devotion, all the inspiration, all the noonday brightness of human genius, are destined to extinction in the vast death of the solar system; and the whole temple of Man's achievement

must inevitably be buried beneath the debris of a universe in ruins – all these things, if not quite beyond dispute, are yet so nearly certain, that no philosophy that rejects them can hope to stand.

(1923, pp. 6–7)

Russell concludes that 'Only within the scaffolding of these truths, only on the firm foundation of unyielding despair, can the soul's habitation be henceforth safely built.' (Russell 1923, p. 7)

These passages spell out an overall worldview derived from science. Despite the eloquence of the two writers, it is unpleasant for many people. Most of us want to be loved, want to know that what we are doing has some significance, perhaps that we will be remembered long after our deaths. To many of us, the 'foundation of unyielding despair' propounded by Kurtz and Russell is unattractive, and we naturally seek for something better. It would not be at all surprising if some people rejected the scientific viewpoint in favour of religious or other beliefs. In part, it seems, this is one of the motivating forces behind creationism. It would also not be surprising if intellectuals took the view, as do some postmodernists, that science is simply one viewpoint among many, and entitled to no special treatment at all.

Let us compare this to another view, one that does not stem from a scientific perspective at all: Frederick Myers also wrote about the general human condition, but he was a proponent of the paranormal. It is interesting to read and absorb his perspective on the human place in the universe, and see the contrast both with that of Kurtz and that of Russell:

A few words will sum up broadly the general situation as it at present seems to me to stand. The dwellers on this earth, themselves spirits, are an object of love and care to spirits higher than they. The most important boon that can possibly be bestowed on them is knowledge as to their position in the universe, the assurance that their existence is a cosmic, and not merely a planetary, a spiritual and not merely a corporeal existence.

and

Perhaps, indeed, in this complex of interpenetrating spirits our own effort is no individual, no transitory thing. That which lies at the root of each of us lies at the root of the Cosmos too. Our struggle is the struggle of the universe itself, and the very Godhead finds fulfilment through our upward-striving souls.

(Myers 1970, p. 422)

Note the contrasts in the views. For Myers, our lives are not doomed to insignificance; spiritual creatures in the universe love and care about us. Our lives are not limited to physical existence on this planet, and what we seek to do here is of profound cosmic significance. In short, the Myers perspective differs completely from those of Kurtz and Russell. Clearly, there will be a strong temptation for every one of us to favour Myers's perspective. It is so much easier to believe that we are important and loved, rather than that we are insignificant and uncared for. It follows that what Kurtz terms the 'transcendental temptation' will always beset humanity. There will always be voices telling us that the perspectives of science are profoundly mistaken, that we matter far more than science tells us, and that we are not alone.

LAUNCHING PAD

This chapter has sought to clarify the nature of science, some of its problems, and its impact on the world. Understanding the nature of science is essential to any study of the paranormal, since both the paranormal and skepticism are defined in terms of science. After some discussion, we produced a simple model of science, describing it as a process of interaction between theory and evidence. Theories spell out ideas about how the natural universe works, and guide the search for evidence. Evidence – of which experiments are the strongest form – in turn decides which theories are accepted, and which are discarded. This circle results in a steady growth of scientific understanding.

We should remember above all that there is nothing mechanical or otherworldly about science. It is a human attempt by clever people to systematically extend our understanding of the natural universe. Since the ultimate jury, the scientific community, is human, it follows that sometimes the jury will be wrong. Since the scientific approach is specifically designed to be checked against empirical evidence, and scientific theories can be, and often are, discarded when the evidence is not supportive, it is likely that scientific knowledge will be reliable. That is, when something is generally accepted by the community of researchers who are expert in the field, they are far more likely to be right than an unsupported, isolated outsider. Science is not infallible, but it is generally very reliable. We will come back to this point when we discuss some of the principles of skepticism. It is important.

We saw that there is nothing formal or fixed about scientific knowledge. At any time there is a 'core' of knowledge which scientists are confident can be regarded as true, and a 'frontier' area, where research still goes on. However, accepted ideas can be scrapped, and discarded ideas can, in principle, return to favour. The entire body of scientific knowledge is constantly changing and reforming itself. In consequence of this, although science cannot prove its theories to be true, the best-established theories are very reliable indeed.

Using these ideas, it was easy to see that creation science cannot be regarded as science at all: its goals are quite different from those of science. Its theory is unalterable, and its proponents' methods of propagation – seeking to have it forced into the core of science – are quite different from those of science.

We also saw that, despite its enormous success, science does have problems. It has difficulty recruiting enough bright young people to continue and expand the scientific endeavour. Many young people may prefer to go into other professions, where the rewards are higher or the training less arduous. In addition, the vast majority of the population have little understanding of the nature of science, and are largely ignorant of its findings. Finally, the scientific picture of the universe, and our place in it, is likely to appear bleak to many people. It relegates our lives, ideals and struggles to insignificance in a vast and impersonal cosmos. Hence, although science is intellectually extremely successful, it faces problems of other natures that may, in the future, possibly threaten its existence.

For an extended glimpse into the workings of science, Le Grand (1990) is excellent. Kuhn (1957) goes through the historic overthrow of the idea that the Sun went round the Earth. In a later book, Kuhn (1966) tries to explain how major change happens in science. An overview of some of the impacts of science appears in Bridgstock et al (1998a).

2 | The paranormal

WHAT EXACTLY IS the paranormal? The term comes from the Greek prefix *para* meaning 'beyond'. Hence, the paranormal is beyond normal. It has been equated with 'weird things', pseudoscience, the supernatural and a great deal more. But what exactly is it?

Carl Sagan takes one approach when he lists the items which can, in his view, be investigated as part of the paranormal. He lists several dozen phenomena, starting with 'astrology, the Bermuda Triangle, "Big Foot" and the Loch Ness Monster' (Sagan 1997, p. 208) and ending with 'innumerable cases of acute credulity by newspapers, magazines and television specials and news programs' (Sagan 1997, p. 209).

Sagan's list is very inclusive, and he also manages to make it quite funny. A problem may occur to the reader, though: why those beliefs in particular? And, if we only use a list like Sagan's, what if other possible forms of paranormality appear? How do we decide what should go on the list? Sagan gives us no guide, no way of deciding what is and what is not paranormal.

Clearly, what we need is some sort of definition. Here is a skeptic's definition of the paranormal. David Marks defines it as follows: 'as related to psychic research, faculties and phenomena that are beyond "normality" in terms of cause and effect as currently understood' (Marks 2000, p. 28).

And here is a believer's definition. JB Rhine is the founder of modern parapsychology. His definition of the paranormal, therefore, should be viewed with a great deal of interest.

Because of their inexplicable nature they are sometimes called 'supernormal' or 'paranormal', indicating that something beyond the normal explanations is required to account for them, but nothing supernatural is implied. Parapsychical phenomena, however rare and mysterious, are natural if they occur at all.

(Rhine 1969, p. 2)

These definitions – along with many others – suggest that the paranormal is concerned with matters strange to our normal thinking, and outside our normal abilities to explain them. The latter, of course, means that they are beyond the current understanding of science. This seems a useful working definition. The parapsychologist Stephen Braude considers a number of possible definitions of the paranormal, and ends up with something quite similar to the definition adopted here (Braude 1979, p. 250).

In essence, despite some serious misprints, Braude seems to state that a phenomenon is paranormal if, and only if, it has three characteristics: it cannot be explained in terms of current scientific theory, it would require major changes to scientific theory to explain it and it thwarts our familiar expectations about what would normally happen.

This is reasonably in line with the definitions we have already looked at. The main difference is that Braude stresses that the anomaly has to be major in scientific terms before a phenomenon can be regarded as paranormal. However, this is a difference of degree rather than meaning. It is reassuring to know that the definition we are adopting has so much in common with another.

It is important to note that all these definitions are negative. That is, the paranormal is defined as being not understood, by both the general population and by science. It follows that the paranormal is likely to be a diverse area. It is easy to find examples of the amazing diversity of paranormal phenomena (Sagan's list of paranormal examples quoted above is a good start). Further, the different claimed phenomena may well actually contradict each other: the claims of some paranormalists will directly contradict that of others, and it is also quite possible that upholders of one belief will regard others with hostility and contempt (Gardner 1957).

Another attribute of the paranormal follows logically from the definition, but is perhaps more surprising. Despite Rhine's optimism, the paranormal must remain forever unknown. It can never become a part of natural knowledge. This is not to limit the possibility of exploration by science: it follows directly from the nature of the definition.

The point is best explained by an example. Consider telepathy – the transference of thoughts from one mind to another, without any intervening physical process. This is clearly part of the paranormal. Now let us imagine an experiment which shows, consistently and repeatably, that telepathy is possible under some circumstances. Furthermore, let us imagine that scientists investigate the matter closely, and are able to explain the telepathy scientifically: perhaps some form of electromagnetic radiation passes from one brain to another, carrying thoughts.

This experiment certainly seems to make telepathy scientifically comprehensible, and establishes that it exists. But it also removes telepathy from the realm of the paranormal. Why? Because according to the definitions, the paranormal consists of that which is unknown and beyond scientific and normal understanding. Telepathy, in our imaginary case, is now established and understood. Far from being paranormal, it is now established as scientific fact, and by that very attribute is no longer unknown. What is more, over time, it is likely that telepathy will become established in the public mind. Indeed, given the sensational nature of telepathy, one can imagine it appearing on the front page of newspapers and the opening announcements of news bulletins. By this process, telepathy would cease to be strange, and would no longer be outside what we regard as normal. Therefore telepathy ceases to be paranormal, and becomes part of the established body of knowledge.

This unexpected result tells us something about the paranormal. It is forever doomed, by its very nature, to remain beyond normal comprehension. No amount of scientific investigation or study can redeem it from this fate. Particular aspects of the paranormal may make their way into the mainstream of knowledge, but the term 'paranormal' will forever refer to phenomena which are not accepted, which are 'weird' and which are deemed to have inadequate evidence to be taken seriously.

As knowledge expands, it is always possible that some aspect of the paranormal will be incorporated into normal science. We may ask whether any aspect of the paranormal has ever made its way into the scientific mainstream? The answer is yes, it has. Of course, the definition of the paranormal depends on the definition of what is known. This happened in the case of meteorites in eighteenth century France. According to the scientific theory of the time, it was impossible for rocks to hurtle through space and strike the Earth. What is more, it is an extremely unusual event, regarded by anyone seeing such impacts as way outside normality. Therefore, it took several decades for the savants of the time to acknowledge that not only was this not contrary to scientific law, but it was happening on a regular basis (Westrum 1978). Of course, impact meteor craters have been found all over the Earth, and it is generally accepted that, at some times in the past, they have radically changed the history of life on this planet (Zanda & Rotaru 2001). What once was scientifically impossible – and hence paranormal – is now part of mainstream science. Similar arguments can also be made for human combustion and firewalking (Kelly 2004). Both have been described as paranormal, but both have been explained scientifically.

This is a strange conclusion. The paranormal as a whole can never be incorporated into mainstream knowledge, but individual parts of it may. I wonder whether the many researchers who seek recognition of their studies of UFOs, crop circles and alternative medicine are aware that, if successful, their area of interest will be transformed into just another area of scientific study, probably with departments in universities studying it, and corporations seeking to commercialise the results.

ANOTHER VIEW OF THE PARANORMAL

A completely different definition of the paranormal is provided by Australian psychologists Michael Thalbourne and Peter Delin (1994). Thalbourne argues that the negative, and broad, definitions outlined above are not useful. Instead, he treats only one part of the larger paranormal field as being truly paranormal – the area of ESP, clairvoyance and psychokinetics – as being what he accepts as paranormal. Thalbourne writes:

> For current purposes, we take it to refer simply to any of three controversial classes of phenomena that are claimed by some to exist: extrasensory perception (ESP), psychokinesis (PK), and life after death. A good many people believe in such phenomena and even claim to have experienced manifestations of them, whereas others, particularly social scientists, reject any notion of this kind as error or wishful thinking.
>
> (Thalbourne & Delin 1994, p. 3)

For an empirical psychologist, it is likely that a simple definition of the paranormal will make research a good deal easier. Thalbourne's concept of the paranormal does not have the bewildering diversity of the other definitions.

On the other hand, what about belief in ghosts, Atlantis, UFOs and creation science? According to Thalbourne's definition these are not paranormal, even though most people would regard them as such. In general, we will keep the field of this study wide, and will use the broad definition for this book. Thalbourne's definition has the virtue of simplicity, and is probably more suitable for research. However, it does not capture the full diversity of paranormal belief, and is therefore less adequate for this book.

WHAT ABOUT RELIGION?

Where does religion fit into this picture? Strictly, religious beliefs are paranormal if we use Braude's definition. Clearly, the actions of a religious deity, or deities, are inexplicable by science. Equally clearly, in our normal experiences we do not encounter God or gods on the way to work, or as we relax in our houses. Most religions have stories of miraculous events such as healing, parting of seas and the destruction of enemies. For example, the Roman Catholic Church must normally have evidence that the person has performed at least two miracles in order to admit someone to sainthood (*National Catholic Reporter* 2003). In short, religion can be regarded as being part of the paranormal, and could be thought of as subject to the same kinds of skeptical investigation as any other paranormal claim.

Should we examine religion with the same skeptical tools that we use on UFOs, telepathy or the lost continent of Atlantis? Opinions within the skeptical movement differ. Paul Kurtz gives an interesting example of the different arguments that can be put forward. Kurtz, however, believes that religion can be examined in precisely the same way as any other paranormal claim. He has done this himself in his book *The Transcendental Temptation* (Kurtz 1986). Although, for purely tactical reasons, Kurtz has taken care to ensure that CSICOP (Committee for the Scientific Investigation of Claims of the Paranormal) is quite distinct from the rationalist, humanist and atheist groups which exist in the US.

Barry Williams, the leading Australian skeptic, has taken a similar stand. He points out that the Skeptics do test some religious claims, and goes on to say that creationism:

> is by no means the only claim made by religious groups on which we do have opinions. Faith healing is one, exorcism is another and there are many more. The connecting factor, and the factor that brings them within the purview of Australian Skeptics, is that these claims are amenable to testing using the tools of science.
>
> (Williams 1994, p. 52)

This is a pragmatic argument, simply arguing that since (as we shall see) the skeptics' aim is to test paranormal claims scientifically, it is pointless to pay any attention to the claims that cannot be tested in this manner. Although Williams does not say so explicitly, this seems to imply that he regards religion as paranormal, but declines to support its testing on the grounds of practicability. As he states at the end of his article:

> Of course, religious beliefs and practices are open to question, as are the selection policies of the Australian Cricket Board or the programming decisions of the Australian Opera, but it is not a role that Australian Skeptics seeks to fulfil. Other organizations do choose to question these beliefs and practices, and it is entirely proper for them to do so in a democratic society.
> (Williams 1994, p. 52)

This pragmatic position is in line with the approach taken in this book. Religious claims are in principle paranormal, but for the most part are impossible to test. What is more, religious beliefs are not dependent on the results of the tests. Indeed, some religious beliefs are essentially a philosophy of life, an attempt to read some meaning and coherence into the universe, and it is hard to see how this can ever be subject to skeptical testing. Thus, although it is perfectly defensible to test paranormal claims which derive from religious beliefs – such as creation science claims, faith healing, weeping and bleeding statues and the legitimacy of claims made about the Shroud of Turin – religious beliefs themselves will be left out of consideration. It is too large an area, and too remote from most skeptical concerns.

PARANORMALITY AND SCIENCE

What about the actual beliefs of paranormalists? There have been many attempts to classify what paranormalists believe (e.g. Palmer 1979). Using mythical approaches and deconstruction, it is probably possible to generate a huge number of different ways of looking at the beliefs. Here, we will look at a simple one: how do the different paranormal fields relate to science? We have seen already that science is a key concept in defining the paranormal, so the attitude taken to science tells us a good deal about what is going on in paranormal fields.

There seem to be several possible approaches that proponents of the paranormal might take toward science. They may seek acceptance within science, or seek to change science in some way. They may be indifferent to science, or even actively hostile to it. We will consider each of these possibilities.

Some parapsychologists actively seek to achieve the status of science. They carry out research on issues like telepathy and precognition, organise research centres and publish the results in properly refereed and referenced journals. The best example of this area is parapsychology, which has explicitly sought to develop in this way following the example of JB Rhine (1969). Parapsychologists, generally, accept that they must put forward convincing evidence for their claims, and so they work their way from the scientific frontier into the core. They have been attempting this for decades. It promises to be a long, arduous journey.

It is clear that parapsychologists' intellectual approach is closely related to the way they organise their work. They aspire to being a part of science, and their organisations are modelled on scientific ones.

Superficially, the second approach resembles the first. It consists of waging a kind of 'guerrilla war' against science, arguing from scientific evidence for the existence of certain kinds of paranormal claim. Creation science is perhaps the most obvious example of this. Creationists publish books and articles purporting to demonstrate scientifically that the process of evolution could not possibly have taken place (e.g. Morris 1984). They also dispute the idea that the Earth is very old, that the continents have moved over enormous periods of time and that the universe originated in an enormous explosion. If one reads creationist books and magazines, which are often referenced science-style, one can emerge with the impression of a huge mass of evidence which is sweeping away the claims of 'evolution-science'. It can be quite impressive.

On the face of it, creation science looks as if it is seeking the status of science. This impression is further reinforced by the claims of creation scientists that their views should receive 'equal time' in schools alongside 'evolution science'. However, appearances are rather deceiving. Creationists are not interested in achieving the status of science in the current meaning of the word. Instead, their goal is to radically change the nature of science, rewriting its basic premises on grounds acceptable to them. In short, creationists do not want to become part of science. They want to take over science and remould it in their chosen image (Kitcher 1982).

What would the new creation-based science look like? As we have seen, creationists have made this perfectly clear. For them, the account of the origins of the world in Genesis is the word of God, and so cannot be wrong. Therefore, logically, whatever theories and findings become part of this new science, they must be compatible with the fundamentalists' interpretation of Genesis (Bridgstock 1986b, p. 81). There is no possibility that any amount of evidence can overturn this view.

In this respect, creation science differs profoundly from normal science. In principle, any part of normal science can be overthrown. The most widely accepted theory or law can be falsified if the evidence against it is sufficiently strong. However, this cannot happen to the creationist view. It is not subject to empirical falsification – at least in the eyes of its supporters.

For this reason, it is difficult to see how creation science can ever be truly scientific. Its dogmatic, fixed nature makes it quite different from the ever-changing, empirically based process of science. However, what the creationists really seem to be after is not gaining scientific acceptance. What they want is to establish their religious views – and theirs alone – as having the same credibility as science. However, they do not wish to do this by the difficult scientific process of formulating and testing theories: they are seeking success through political pressure. For this reason, creationism differs fundamentally from science, and it also differs profoundly from parapsychology. For the most part, parapsychology seeks to adopt the methods of science, and looks forward to being accepted within the scientific community. By contrast, creationists disagree profoundly with the actual goals of science, and seek to replace it with their own approach. As far as we can tell, therefore, parapsychology and creation science differ profoundly in their attitudes to science. The former seeks to become scientific, the latter actually wants to destroy science as it is currently practised and replace it with a different model.

Another way is to ignore science altogether, and to adopt an 'anything goes' approach. Perhaps the best-known example of this approach is the philosophy of the 'New Age'. This is a movement composed of many different elements so that, as far as anyone can tell, there are almost no commonly held beliefs at all. Instead, each individual is urged to choose from a range of beliefs and practices, whatever works for them.

Cultural commentator Lisa Aldred summarises it like this:

> The term "New Age" is often used to refer to a movement which emerged in the eighties. Its adherents ascribe to an eclectic amalgamation of beliefs and practices, often hybridized from various cultures. New Agers tend to focus on what they refer to as personal transformation and spiritual growth. Many of them envision a literal "New Age" . . . it will be achieved through individual personal transformation
>
> (Aldred 2002, pp. 61–2)

In this context, each individual must seek their own path to spiritual enlightenment and personal growth. Therefore, New Age adherents use a combination of almost any elements that work for them. There may be parts of Eastern religions, astrology, clairvoyance, pendulums, crystals, pyramids and almost anything else.

If questioned about science, New Agers are likely to argue that science is only one way of knowing, and that it has profound limitations which prevent its grasping far more important truths. In one sense, of course, this is completely true. Science has not been successful in some areas of study. However, there is a difference between being aware of science's limitations and repudiating sciesence altogether, and the New Age does not seem to incorporate this insight.

How is the New Age organised? It appears to be what Goode (2000) would call a 'grassroots' organisation. That is, it does not have a centre, but exists around loose networks of believers with a wide variety of beliefs and organisational forms. Of course, it is hard to imagine such a loose set of beliefs being based around any other kind of organisation: there is no centre, no fixed dogma, and no one with either the power or the authority to impose a particular belief.

Similar considerations apply to the UFO movement. Some ufologists seek academic respectability, others are hostile to the idea of science. Some are interested in unexplained lights in the sky, others believe that they have been abducted and sexually violated by aliens on a regular basis. In short, although the content of the beliefs of New Agers and ufologists differs widely, they both reflect the decentralised, grassroots nature of the organisation in that they are extremely diverse and variegated.

We might make a few quick points about the New Age before proceeding further. First, it contrasts strongly with both parapsychology and creation science. Parapsychology seeks, much of the time, to imitate the rigorous methods of science, while creationists argue that the Bible, as the word of God, must receive literal acceptance. By contrast, the New Age has no particular set of beliefs. In one way, this can make the New Age most attractive. People are encouraged to seek growth in whichever way suits them, making New Agers generally very tolerant and easygoing. In addition, since each person can choose the beliefs or combinations of beliefs that suit them, the New Age has almost no boundaries. It can become extremely popular with people who have almost nothing in common with each other.

At the same time, the New Age, and similar beliefs, have profound limitations. If the truth, for each individual, is what serves that particular person, then there seems to be no way of embarking on a common project to find out how the universe works, or what is the best way to cure disease or produce wealth. What works for one person may not work for another, leading to a morass of individual insights, but no means of separating truth from falsehood.

These are some approaches that paranormalists take to science. Some are profoundly hostile to science. Of course, this takes them completely out of the realm of discussion, as they do not accept evidence-based approaches in science. There are almost certainly others. However, the approaches we have looked at here are probably the most important, and we will return to them later.

WHY DO PEOPLE BELIEVE IN THE PARANORMAL?

We have seen from the definitions of the paranormal that these phenomena cannot be properly supported by evidence. As soon as they are supported in this manner, they cease to be paranormal in nature. So the question remains: if we grant that most paranormal beliefs are insufficiently supported by evidence, why is belief so widespread?

Psychologists and philosophers have spent a good deal of effort trying to answer this question. Answers have come from all directions. It seems fair to say that we do not clearly understand the prevalence of paranormal belief. However, the reason is not that theories of paranormal belief have not been substantiated, but because so many have received support. In my view, so many theories have been shown to be plausible that we need to start working out which are the most important. Since evidence on this is lacking, we will simply review some major theories, to give an idea of their diversity. It is certain, a lifetime can be spent on this question!

One set of explanations comes from philosophy and logic. We humans have limited powers of reasoning, and we also have limited information. As we have already seen, this means that science constantly has to revise its theories. In the same way, humans often have ideas about reality that can be wrong, and need to be revised. In some cases, these ideas lead to paranormal belief (Shermer 1997, pp. 55–8).

Let us look at a simple example. You have an illness which normal medical science cannot cure. Perhaps it is a persistent viral infection, or aches and pains that nothing can permanently relieve. You go to a homoeopath, or a psychic healer, feeling doubtful. To your amazement, the next day your symptoms are gone. You are cured! It is perfectly natural to attribute this to the healer. As we shall see in Chapter seven, this is probably not the case, but it seems reasonable to become a believer in alternative medicine.

Let us look at a more lurid example. Psychologist Susan Clancy set out to examine the phenomenon of UFO abduction. These are the astonishing cases where people claim to have been taken aboard an alien spacecraft and – often – subjected to strange, sexually related tests and procedures. Clancy set out to study people who have 'memories' of these strange events. Contrary to popular belief, she did not find abductees were at all mad. Instead, she concluded that 'the reason they ultimately endorse abduction is actually quite scientific: it is the best fit for their data – their personal experiences.' (Clancy 2005, p. 52)

What is Clancy talking about? What personal experience can lead people to conclude they have been abducted? One of the main experiences is sleep paralysis, in which we wake from sleep but our bodies have not yet acquired the ability to move. Here is how Clancy describes an episode of waking up paralysed:

> You're wide awake. You open your eyes and try to get up, but you can't. You're on your back, completely paralysed. There is a sinister presence in the room. Then you hear something. Footsteps? Something is padding softly through the room. Your heart is pounding and you try to scream but you can't. Whatever is in the room moves closer. Then it's on top of you, crushing your chest. You glance to the left and see small shadowy forms in the corner... The experience passes... What the hell just happened?
> (Clancy 2005, pp. 48–9)

This terrifying experience, apparently, happens to about 20 per cent of the population. Some people – a minority of those who experience it – finally decide that the only explanation is that they have been abducted by aliens. Given the terrifying nature of the experience, this is at least understandable. However, there is a perfectly good scientific explanation, it's simply that the abductees do not know what it is.

These two examples show how perfectly normal, sane people can adopt paranormal beliefs as an explanation of their own experiences. We can say with some assurance that they are wrong, but there is nothing mad about their views. They have fallen into traps that are common to all humanity, and imposed wrong explanations on insufficient data. An analogous process can occur to people who are devotees of alternative medicine. Many people have bad experiences with mainstream medicine. Doctors can be impersonal, rude or just plain wrong. Since sick people are extremely vulnerable, this can cause great suffering. In these circumstances, some people may conclude that alternative medicine offers better chances of a cure.

This problem is compounded by a second human tendency. Again, it is normal and natural, but it causes problems. We often see patterns when there are none. Indeed, it is very easy to watch the mind at work creating patterns. Flip a coin a dozen times, and note down the sequence of heads and tails which emerges. If you are anything like me, as you are noting the results, you will find yourself looking for patterns. You may find yourself thinking: 'Two heads, one tail, so it should be two heads next ...', and so on. It takes a distinct effort to stop doing this. Many optical illusions take advantage of this, inducing us to see things that are not there. Good evolutionary reasons have been advanced for this tendency. After all, in the early days of humanity it was essential to see patterns quickly. Conclusions like 'That's food, we can eat it', or 'There's a predator in the grass!' need to be made quickly in the name of survival. It is more important to see a pattern quickly than to be infallibly right.

This pattern-seeing tendency can lead us to see patterns when there are none. As we shall see in Chapter five, we may experience an amazing coincidence and conclude that some paranormal mechanism is at work. In reality, it can simply be chance. So this normal – and necessary – human tendency can also lead to erroneous conclusions.

There is a third perfectly normal human tendency which also leads us toward paranormal beliefs, and we have already encountered it. David Hume, the great philosopher, points to the agreeable emotions of 'surprise and wonder, arising from miracles' and argues that this 'gives a sensible tendency towards the belief of those events, from which it is derived' (Hume 1966a [1777], p. 117). This is not quite universal, but many people are tempted in exactly this way. I remember as a child terrifying my contemporaries by making up ghost stories. I loved being the centre of attention, and they loved being terrified. Later in life I made a simple attempt to understand and classify this type of appeal (Bridgstock 1989). To my knowledge, little research has been done on this phenomenon, but it does predispose a large fraction of humanity to belief that 'something is going on' beyond the visible, natural world. And, as Hume says, if we gain satisfaction both from proclaiming and hearing about strange events, then it is hardly surprising that whole industries exist devoted to the paranormal.

Finally, there is the vexed area of abnormal psychology. It is often thought that paranormal believers are 'nuts', and indeed it is suggested that people with lower intelligence or mental problems are more likely to hold paranormal beliefs, and to report that they have had paranormal experiences. Other research indicates that this is not so: the claims are not well established (Smith et al 1998). However, there are many other effects as well. Michael Thalbourne and Peter Delin (1994), Australian psychologists, have devised the term 'transliminality'. This refers to people – not mentally ill by any means – who are far more in touch with the unconscious mind than the rest of us. For some people, notably artists, this may be a positive asset, but it may mean that the boundaries between reality and fantasy are somewhat blurred, and so paranormal belief and experience may be more accepted.

There are a large number of human traits and dispositions which strongly prejudice us towards accepting paranormal explanations and propositions. For the most part, these are not in any sense bad or unhealthy: they are simply a part of our human make-up. It may well be that some of them, in some circumstances, can be beneficial. For these reasons, it is not at all surprising that paranormal belief is strongly prevalent today. This is even more so when, as we will see in the next section, there are social configurations in society which support paranormal beliefs, and help to protect them from critical scrutiny.

THE PARANORMAL AND SOCIETY

One of the few serious studies of the social organisation of the paranormal has been carried out by sociologist Erich Goode (2000). Goode made a reputation as a student of 'social deviance' (conduct contrary to that expected by society) before beginning to investigate the paranormal. In his major work on this field, he distinguished five different types of paranormal organization, and suggests that it is worthwhile studying the links between these organizations and the types of knowledge they claim to possess.

Goode's classification of paranormal organisations appears in Table 2.1. As we will see, his typology appears both incomplete and unsatisfactory.

Table 2.1. Types of paranormal organisation, with examples, according to Goode (2000)

Social Isolates – cranks
Client-Practitioners – clairvoyants, astrologers, alternative healers
Religious beliefs – creation scientists
Pseudo-scientists (or proto-scientists) – parapsychologists
Grass-roots – UFOlogists

However, it is an important pioneering effort, and deserves to be taken seriously. After we have looked at Goode's list of paranormal organisations, we will consider some problems with the ideas, and add some items to his list.

THE CRANK, OR SOCIAL ISOLATE

Goode's first type of paranormal organisation is not really an organisation at all. It consists of the socially isolated paranormalist, often referred to as the 'crank'. The crank does not really 'come alive' in Goode's discussion. However the British astronomer Patrick Moore (1974) has written a delightful book in which he tries to understand how these social isolates see the world. Moore does not endorse the views of the 'independent thinkers' as he calls them, but he is quite sympathetic and understanding of their outlook.

Goode distinguishes the crank sharply from the revolutionary scientist, who alters the theories of science with brilliant new ideas, such as Galileo and Einstein (Goode 2000, p. 83). This is an important point, since many cranks argue strongly that they have new ideas, at least as good as Einstein's, and are being persecuted like Galileo. Einstein certainly revolutionised science with astonishing new ideas, Goode agrees, but he was not an isolate. He was in contact with many other scientists, and submitted his papers to scientific journals for consideration. This is in contrast to the cranks, who are usually isolated from mainstream science and make few or no attempts to publish their ideas in the normal scientific publications.

Cranks, suggests Goode, have a profoundly ambivalent attitude towards science. They believe strongly in the great importance of their theories, and tend to be rather humourless on anything to do with them. Therefore, they aspire to the recognition and significance to which, they believe, their ideas entitle them. But also, they are profoundly antagonistic to the science which rejects or ignores them, and takes no account of their ideas.

In Goode's view, the crank is a grossly understudied phenomenon. It would certainly be of interest to understand these people better. They toil away for years, developing and writing theories, and sometimes inventing intricate devices – such as anti-gravity or perpetual motion machines – which they believe will revolutionise human life. How do such people maintain their faith in the absence of recognition or indeed notice from anyone of any significance? It seems likely that such a study would be rather sad. One suspects that studies of the crank will be studies of loneliness, of self-delusion and of ultimate failure. Still, Moore's book is kind and, on the whole, hopeful. A chapter in the autobiography of Arthur Koestler (1969) also gives some insight into this area – Koestler's father seems to have been a crank.

THE CLIENT-PRACTITIONER

Goode's second category of paranormalist – the client-practitioner – does not show any of the characteristics of cranks. These people provide a skilled service to a client in exchange for a fee. Examples include astrologers, clairvoyants, homoeopaths, chiropractors and naturopaths. In this way, they are modelled on prestigious professionals in society, such as doctors or lawyers. Indeed, with the rising interest in some forms of alternative medicine, some medical doctors actually are naturopaths or homoeopaths.

Goode makes an important point when he observes that client-practitioners are not isolates (Goode 2000, pp. 96–7). They draw on a tradition of paranormal knowledge which might be quite old. Homoeopathy is nearly two hundred years old, and astrology can be traced back at least five thousand years. In addition, they form networks of practitioners, similar to those in medicine or law, and often congregate in learned societies to discuss developments in their fields.

Another difference is that the client-practitioner, like the doctor, is not a theorist. The practitioners are not interested in producing theories that will revolutionise science. They are interested in techniques that will help the people who come to them, and whose money they receive. Therefore, it is likely that their knowledge will be applied in nature: they have a good idea of what works and what does not.

Goode makes a poignant point when he argues that when a member of the public consults a practising paranormalist – astrology, palmist, homoeopath or whatever – both participants have a strong need to believe that something real and valid is taking place. The practitioner has often invested years of study and training in the discipline, and will be unlikely to abandon it lightly. The client usually has some matter that is troubling them – health, emotional or personal problems, or uncertainty about the future – and needs reassurance and guidance about their troubles. It follows that not very much credence can be given to 'customer' testimonies about the value of paranormal practitioners. Indeed, when skeptic Karen Stollznow of the Australian Skeptics visited paranormal practitioners, it appeared that they were unable to make any meaningful diagnoses of the illness, or to offer any useful information (e.g. Edwards and Stollznow 1998).

RELIGIOUS-BASED PARANORMALITY

Goode's third category consists of those paranormalists whose beliefs are linked to some form of organised religion. Although religious beliefs themselves are not relevant to this study, their paranormal manifestations are. Probably the most dramatic of these paranormal beliefs, in recent years, has been the eruption of creation science as a serious rival to ordinary science. The creation scientists' beliefs are directly related to the beliefs of fundamentalist Christians. Indeed, as we saw in Chapter two, there is a quite direct link (Bridgstock 1986b). Creationists are required to believe in the literal truth of the Bible, and also argue that their beliefs are scientifically verifiable. However, there are other links between religious belief and paranormal claims. Roman Catholics are often receptive to claims of weeping, bleeding and moving statues of the Virgin Mary. Charismatic Christians often claim to be able to heal illnesses through religious faith. In addition, the author has been told of Islamic attempts to explain natural phenomena by means of 'djinns'. In short, religious beliefs are clearly a source of, and of support for, paranormal claims.

The reasons why paranormal claims should find support among religious communities is quite obvious. Some paranormal claims seem to provide evidence that supports some religious beliefs. Therefore, claims of paranormal phenomena often find a ready hearing – and financial contributions as well – in some religious quarters.

There are dangers both for paranormalists and religious people in becoming too closely associated. The dangers for religious people are obvious: what happens if the paranormal claims are shown to be bogus, or mistaken? At the very least there will be embarrassment, and there may be a substantial loss of faith among some religious people. The dangers for the paranormalists are that, by associating themselves with a particular religious viewpoint, they may alienate people with other faiths. Thus, the hard-line Biblical fundamentalism displayed by creation scientists clearly alienates less fundamentalist Christians. In the same way, the recurrent claims of weeping and bleeding statues arouse a good deal of amusement and perhaps contempt among people not of the Catholic faith.

Despite its dangers, association with a religion can confer major benefits upon a paranormalist movement. It can provide a major audience, a continuing source of funds and – if used adroitly – a political base for the acquisition of major governmental support.

PSEUDOSCIENCE OR PROTOSCIENCE

Goode's fourth category is that of the pseudoscientist, or protoscientist. We can define this as a group who claim to be scientific, who seek to use scientific methods, but whose claims are not accepted by mainstream science. Many groups may make some sort of claim to scientific respectability to enhance the plausibility of their beliefs. Parapsychologists, beginning with JB Rhine, have made a sustained attempt to be accepted scientifically. A less impressive example of this is the crop circles enthusiasts' claims to be called 'cereologists', their use of scientific instruments and terminology, and their holding of scientific-style conferences (Schnabel 1993).

The embracing of a scientific approach (or what they understand to be scientific) by some paranormal believers has some clear advantages for them. It may be possible to gain funding for research and perhaps a place within the university system – as was the case with parapsychology. And there is no doubt that a few indications of scientific acceptance go a long way in the paranormal community. For example, with a good deal of hesitation, the journal *Nature* published a paper by two parapsychologists claiming experimental success in promoting remote viewing (Targ and Puthoff 1974). The paper was accompanied by an editorial expressing doubts about the validity of the paper's findings (*Nature* 1974), but the authors gained great prestige from the publication. As Randi puts it, in a scathing review of the entire episode, 'the two authors were the toast of the psi world, being asked to speak and opine on all aspects of the paranormal' (Randi 1982, p. 131). Being elevated to guru status is a fine reward indeed.

As we have seen, no scientific findings are ever completely immune from being questioned and falsified, and so claims that cannot be reliably backed by evidence are liable to be discarded. This is exactly what happened to Targ and Puthoff's claims about remote viewing. Two New Zealand psychologists attempted to replicate Targ and Puthoff's findings (Marks and Kammann 1978; Marks 2000). They failed, and produced a good deal of evidence that the original experiments were fatally flawed, and could not be taken seriously.

The position of parapsychology is even more insecure than this suggests. Paranormal findings that cannot be replicated will be discarded from what science regards as true. Although, as we have seen, if a paranormal finding is verified and supported by further evidence, it then ceases to be paranormal. It simply becomes part of scientific knowledge, and loses its paranormal qualities. In the long run, it is likely that the field will either produce solid, replicable findings and become part of established science, or it will gradually abandon its attempts to join the sciences, and become more and more 'way out'. It is also possible that it will simply continue in this indeterminate grey area, making claims, producing occasional results, but convincing no one in mainstream science. This would be a sad, but likely outcome.

GRASSROOTS OR SOCIAL MOVEMENTS

Good's fifth and final category is that of the grassroots movement. These are often termed social movements by sociologists, and are rather a residual category. They have an active membership – often a very large one – but are not affiliated with a religion, nor are they particularly interested in mimicking mainstream science. Goode gives the example of the ufologists as being a grassroots movement. Certainly, the UFO movement appears to have little support from any source apart from the commitment of its own members. Goode writes:

> They are sustained less by individual theorists, a religious tradition or orga-
> nization, a client-practitioner relationship, or a core of researchers than by
> a broad-based public. In spite of the fact that it is strongly influenced by
> media reports and the fact that there are numerous UFO organizations and
> journals, the belief that unidentified flying objects (UFOs) are "something
> real" has owed its existence primarily to a more-or-less spontaneous feel-
> ing among the population at large. The ufologist's social constituency is
> primarily other ufologists, secondarily the society as a whole.
>
> (Goode 2000, p. 82)

Goode does not follow up his ideas about the UFO movement very far, but a few ideas do seem to follow quite directly. First, the movement is heterogeneous. One wing is devoted to the 'nuts-and-bolts' idea that UFOs are alien spacecraft, while others are convinced that they are something much stranger, perhaps from another dimension or time. Some activists are rather skeptical, debunking some cases, while others are speculative, spinning entire cosmologies from their ideas about what UFOs may be (Randles, Roberts & Clarke 2000). More recently, a rift has developed between those who are strong believers in UFO abductions and those who are skeptical about this idea. The UFO field is studded with organisations referred to by impressive sounding acronyms, and there seems to be much internal conflict and suspicion between the different groupings. There is also a large number of conspiracy theorists within the movement, who believe that large-scale government cover-ups are occurring, and that the people are entitled to be told the 'truth' (e.g. Good 1987).

This sort of structure has some advantages, particularly in the modern age of the internet. Almost anyone with any type of UFO belief can find someone – and probably whole groups of people – who share their views. Thus, the UFO movement can easily recruit supporters. At the same time, the diversity and incoherence of the movement means that the more sober believers are generally categorised with the more 'way-out' ones, and the entire movement is regarded as very strange. Therefore, in the absence of conclusive proof of the existence of UFOs, it really seems as though UFO believers are forever doomed to be outsiders.

Of course, this will rapidly change if clear and conclusive evidence of aliens is discovered. Once the first little green (or grey) alien addresses the United Nations, or the first unambiguous and indisputable sighting is made, UFOs will become firmly established as a field of study. At the same time, precisely because of this, UFO beliefs will cease to have any paranormal features, and the current believers are likely to be swept aside.

SOME FURTHER THOUGHTS ON GOODE'S IDEAS

What should we make of Goode's ideas about paranormal organisations? On the one hand they are intriguing, and they do seem to throw light on the way that paranormal beliefs are 'rooted' in our society. Paranormal claims do not float about in thin air, but come from people in specific contexts, supported by specific organisations. On the other hand, there are problems. First, of course, Goode does not actually show that the type of organisation affects the type of paranormal belief. It is simply, at this time, a speculation that needs to be followed up.

In addition, Goode's list of types of paranormal belief organisations is incomplete. There would appear to be at least two other types of organisation which can sustain paranormal beliefs, and we will deal with them here. These are cults, and the movements which have been loosely called 'witch-hunts', though they can also be termed 'moral panics'. Neither of these resemble any of Goode's types of organisation but each has been shown to support paranormal beliefs.

We will look at cults first. The term 'cult' is itself pretty unfavourable, and conjures up visions of fanatical groups of people with weird ideas. However, this is not their key feature. Margaret Thaler Singer and Jana Lalich (1995) define cults briefly as follows:

> It denotes a group that forms around a person who claims he or she has a special mission or knowledge, which will be shared with those who turn over most of their decision making to that self-appointed leader.
>
> (Singer and Lalich 1995, p. xx)

Singer and Lalich stress that all cults are not harmful. They believe there are two main categories, the cults which gain a great deal of control over their members' lives and the cults which are essentially commercial in nature. They – and we – focus on the first of these. However, the key point has been spelt out in the definition above. In a cult, as defined in this way, the views of the leader are paramount. The followers have in large measure abdicated their own judgement in favour of that of the leader. Therefore, whatever the leader regards as being important will be important to the cult. These important matters may or may not be paranormal in nature. In Charles Manson's murderous cult, known as 'The Family', for instance, the leader's main concerns were political, and he sent his adherents out to murder people with the intention of creating a race war (Bugliosi 1999; Sanders 2002). Michael Shermer (1997) has written about one of the strangest cults of all, Ayn Rand's 'Objectivist' movement. This movement was explicitly rationalist and anti-religious and anti-paranormal. The key ideas of objectivism were that a real world exists, and that we must use our rational intelligences to work out how it works, how we should behave in it, and also that we should take responsibility for our actions. However, Shermer argues, despite its avowed objectives, Rand's movement had the key characteristic of regarding Rand's views as being infallibly correct, and so fell into the definition of a cult. Despite her stress upon rationality, Ayn Rand ended up as a cult leader.

When the cult leader's beliefs involve the paranormal, so will the cult's. This can have terrible consequences. For example, dozens of members of the Heaven's Gate Cult had themselves castrated, and later committed suicide in the belief that the flying saucers were coming to rescue them (Miller 1997). In a similar tragic vein, members of the 'breatharian' cult starved themselves to death in the belief that, having reached a certain level of spiritual awakening, they no longer needed food (Anonymous 1999).

There seems to be a good case for including the cult as one type of organisation which can support paranormal beliefs, and which is not included in Goode's list. The last possibility – the last way in which paranormal beliefs are rooted in the larger society – is not, properly speaking, an organisation. Instead, it is a terrifying social movement.

THE WITCH-HUNT OR MORAL PANIC

Finally, one other social formation which might support paranormal beliefs is perhaps the most bizarre and frightening of all. This is the 'witch-hunt'. The key idea here is that some beliefs acquire an explosive, infectious force, in which more and more people fall prey to a fear-filled belief and infect others with the same ideas. The classic example of this was the witch-hunt of the fifteenth and sixteenth centuries. Historians have spent a good deal of effort working out why this movement took place (e.g. Trevor-Roper 1969). Michael Shermer (1997) has written about the sources of witch-hunts. Essentially, the key element is a state of fear and pervasive uncertainty. When the first accusations are made, they often involve a matter about which many people feel both afraid and defenceless, so that wild accusations, possibly by a mentally unstable person, are given credence. As more and more people succumb to the hysteria, the witch-hunt gathers way. Many people may be killed or persecuted. However, the mechanism of the witch-hunt is inherently unstable, and eventually there will be a decline or a collapse in the movement.

Witch-hunts, like cults, may or may not involve the paranormal. The great early witch-hunts certainly did. They postulated a vast secret movement of witches with unearthly powers, often copulating with Satan and responsible for great amounts of suffering in the community. More recent witch-hunts and panics have varied. There was certainly a wave of panic and persecution over accusations of child sexual molestation – the recovered memory movement – and the alleged satanic mistreatment of children in childcare facilities (Nathan & Snedeker 1995), but these were not paranormal in nature. A survey of panics shows that there is a surprising number of them: some are paranormal and some are not (Bartholomew and Goode 2000). Recently, Goode himself has argued that moral panics may be important in skeptical work (Goode 2008).

The witch-hunt, for a little while at least, seems to be capable of supporting paranormal beliefs and may, in some circumstances, make them a paramount feature of social life. For these reasons, it is appropriate to add it to our list of social formations supporting the paranormal.

HANDLING CRITICISM

What does this survey of possible social formations supporting the paranormal tell us? One point seems to stand out. None of the formations need be confined to supporting paranormal beliefs. They can also support, for example, dissident political beliefs, or methods of counselling which differ from the norm. In short, paranormal beliefs are founded in the sort of organisations and processes which exist in society all around us, and which can be beneficial. This is an important point, because although the paranormal may seem to be 'out of this world', the ways in which it fits into society are in fact very mundane and normal. Whatever the truth of paranormal claims, the people who make such claims are human beings.

We might also note that the paranormal can be said to be 'shielded' from critical inquiry by the social formations in which they exist. As we have seen, science is relentlessly expanding the areas of phenomena that it seeks to explain, and also regards the claims it has already accepted as being open to doubt. We also saw Carl Sagan explaining exactly how merciless scientific criticism can be. Victor Stenger, a particle physicist, attempted to bring this sort of criticism to bear on paranormal claims, and was quite disconcerted by the results.

> When I began applying to claims of psychic phenomena the type of hard-nosed show-me criticism I was accustomed to both giving and taking as a particle physicist, I suddenly found myself subjected to personal attack. I began to see my name in print, not just as another coauthor of a scientific article, but denounced as a "witch-hunter", "McCarthyite", and even as someone possessed by the Devil!
>
> (Stenger 1990, p. 54)

Paranormal proponents seem to have difficulty withstanding scientific criticism, and it is logical that they would seek to avoid it. Therefore, the question is, how do beliefs that are rejected by science sustain themselves against this relentless pressure? Each type of social formation supporting paranormal belief has ways of protecting itself against scientific pressure. The only exceptions to this are, perhaps, the protoscientific communities. This point is so important that it is worth going through each of the types of formation, and seeing exactly what the shielding is.

In the case of cranks, why do these claimants not face scientific inquiry, and not have to answer awkward questions about the quality of their evidence? The answer can be found in their very isolation. Because the crank is not part of a critically inquiring community, the types of questions and objections which scientists routinely raise regarding other scientists' work do not apply to the crank. The latter may berate scientists for being closed-minded, but isolation is precisely what enables these strange beliefs to persist.

We have already seen that the client-practitioners are also shielded from the rigour of critical inquiry. They learn their craft within a context of belief, work within a network of similarly minded practitioners, and face clients who, as we have seen, have a strong commitment to believing in what they are told. The client-practitioners, therefore, like the cranks, are shielded from critical inquiry – unless a skeptic like Karen Stollznow makes her way into their offices.

Paranormal claimants who are associated with a religious community also have a strong shield against critical inquiry. For the most part, they will be presenting their views to people who share their beliefs. In addition, people questioning those beliefs can be regarded as heretics or blasphemers. I have faced this with creation scientists. It is hard to discuss the merits of creation science with fundamentalist Christians, since they equate any criticism of the claims as an attack on Christianity itself, and they cannot see the difference.

As Goode points out, grassroots paranormalists, people like ufologists, primarily address their claims and views to other ufologists. In consequence, skeptics and critics are likely to be regarded as outsiders. There are quite savage disputes among ufologists, but in the main these are over the finer points of ufology, rather than over the question of whether UFOs exist at all. Often, UFO theorists move towards a conspiracy theory about why their beliefs are not accepted. These theories have been given florid form in the *Men in Black* films. Although these films are very funny, the reality they suggest is terrifying. What is more, belief in conspiracies can lead ufologists to be profoundly suspicious of criticism; the critics' motives may be suspected.

Cults, of course, often shield themselves from critical inquiry by their organisation. They isolate their members from the larger community and usually subject them to relentless social pressures to conform. In some cases, the isolated people within these organisations can go for years without encountering anyone who has anything critical to say about the organisation. Anti-cult campaigners such as Hassan (2000) argue that one of the most effective ways of freeing a person from a cult is to expose them to trusted people who do not accept their beliefs, or who have accepted them but now believe are wrong. However, in the short term at least, the cult has strong defences against the criticisms of the outer world. Most cult members will not hear such criticisms. If they hear them, they will not listen to them, and if they listen, they will probably not believe what they are told. Thus, cult members are largely insulated from outside criticism.

The last type of social formation supporting paranormal beliefs, the witch-hunt, appears rather different. It often has little organisation. How then can we say that the beliefs are insulated from criticism? The answer is that in the long run they are not. However, as we have seen, the witch-hunt relies upon a pervasive unease in society at large. There is a general feeling that something is terribly wrong. Thus, when the witch-hunt gathers momentum, people seeking to criticise it face the accusation that they are not concerned about the manifest evil that is being revealed. Indeed, they may themselves become part of the target population. In general, this cannot last but, in the short run at least, the powerful emotions unleashed by the hysteria of the witch-hunt almost guarantee that voices seeking to criticise the campaigners will be silenced – perhaps permanently.

It seems clear, that the social forms that support paranormal belief do so, at least in part, by insulating believers from the types of criticism that are routine within science. However, we cannot conclude from this that the paranormalists are always wrong. Perhaps sometimes they are right, and science is wrong. Looking at the organisations which support paranormality is not to say that paranormalists are deluded, only that being human beings, they belong to specific organisations which are congenial to themselves and their beliefs.

LAUNCHING PAD

We have come a long way in this chapter, and it is worth trying to pull it all together. We have started with an attempt to define the paranormal, and found that a double-barrelled definition would be most useful: the paranormal is *both* contrary to science *and* strange to our everyday experience. We also saw that this type of definition has a number of strange consequences, and we have spent some time spelling them out.

First, from the definition it follows that not only is the paranormal unknown, it can never be known. After all, if some paranormal ability or phenomenon becomes scientifically credible – such as an explanation for telepathy – then it follows that it is no longer paranormal. It seems that a number of phenomena have followed this course: these include human combustion, meteorites and fire walking. That is, they have all been regarded as contrary to scientific logic and normal experience, but have been explained scientifically and are now largely accepted.

Another consequence of the definition is that the paranormal is likely to be extremely diverse. From the definition, anything that is contrary to science and contrary to normal experience is paranormal: the phenomena need have nothing else in common at all. And, indeed, when we look at the paranormal, it is a very mixed bag. In fact, it would be perfectly possible for different paranormal claims to directly contradict each other. And indeed, we do find this. For example creation science posits that the world is only about ten thousand years old, while believers in Atlantis argue that this ancient civilisation was flourishing ten thousand years before Plato – that is, about 12 500 years ago. Clearly, they cannot both be right.

We might also note some distinctly odd phenomena which are not strictly paranormal. Strange forms of life, such as the Loch Ness Monster, Bigfoot and the Yowie are not included. There is nothing paranormal about the existence of unknown forms of life. Indeed, we keep discovering such forms, including some that have been though to be extinct. Therefore, the Loch Ness Monster, though strange and unusual, does not fall into the paranormal category (although it might be possible to argue that, since the Loch has been studied and surveyed for so long, any monster residing therein would probably have to have evaded detection by paranormal means).

UFOs are a more difficult case. Strictly speaking, there is nothing paranormal about the possibility that more advanced civilisations than ours exist in the universe, and that some of them might have paid us a visit. Therefore the older, nuts and bolts ideas of UFOs do not fall into the paranormal category. On the other hand, some manifestations of UFO claims do clearly fall into the paranormal. The spacecraft apparently appear and disappear at will. UFO abductees relate how the aliens could paralyse and control them by some kind of mind force, and some relate how they were dragged through a windscreen without, apparently, damaging it at all (Jacobs 1998). These can reasonably be classed as paranormal.

We saw that there is a good deal of argument among skeptics about whether religion should be included in the paranormal. It does meet the definition, but many skeptics leave religion as a separate field. The exception is where the religious beliefs are clearly involved in some checkable claim. For example, the creation scientists claim that there is scientific evidence for their beliefs; many religious people believe in weeping or bleeding statues or miraculous cures. Skeptics can research all of this, though not usually the religious beliefs themselves.

We also saw that paranormal believers differ in what they are trying to do. Some – like the parapsychologists – are striving to achieve the status of science by adopting the scientific method. Others carry out guerrilla warfare against science, seeking to replace it with methods of their own. Yet others are indifferent to science, or hostile to it.

We also looked at sociologist Erich Goode's classification of the types of social structures that support paranormal beliefs. This was a fascinating tour of the world of cranks, client-practitioners and other social forms. One of the most intriguing types of movement is the grassroots organisation, such as that for ufology. However, it does look as if Goode has not listed all social formations, and we looked at paranormal beliefs supported by cult organisations, and the witch-hunt phenomena which can, in the short run, produce terrifying storms of fear and persecution. It also looks as if the types of knowledge claimed by different types of paranormal organisation are related to the organisation's characteristics – so the New Age is decentralised, and so on – but this is such a new field that we cannot say much about it.

We now have some idea of what the paranormal encompasses, and what shape it takes in modern societies. It is clear that paranormal beliefs can be traced back through human civilisation. Equally, though, a movement has arisen which is usually strongly critical of paranormal claims. Modern skepticism has ancient roots as well, and in Chapter three we will see what they are.

The paranormal is all around us. If we go into any bookshop, unless the owner is a militant non-believer, we will find at least one paranormal section. Perhaps there are two main points to grasp. One is the sheer diversity of the paranormal, the other is the intellectual problem of grasping exactly what the paranormal is – since it is defined in terms of what it is not.

For the former, the 'weirdness sampler' in the first chapter of Schick and Vaughn's (2002) book is a reasonable place to start, as is the sample in the first couple of chapters of the book by Erich Goode (2000).

For a grasp of the problem in defining the paranormal, the series of definitions offered by Braude (1979) is useful. Bear in mind, though, that his final definition is wrongly stated (I think because of poor proofreading) and a corrected version appears in this book. For what this means in practice, have a look at Blackmore's account of the problems that she had in researching a concept which seemed to have no substance (Blackmore 1996). She spent years of her life pursuing psi (a key paranormal concept) before concluding it was hopeless.

3 | Skepticism – from Socrates to Hume

W E WILL now look at the third term in the book's title: skepticism. As with science and the paranormal, we will begin with the intellectual aspects of skepticism. Then, just as we did with science and the paranormal, we will ask how skepticism is rooted in modern society. Because there is a long history involved, we will look at some important historical figures here, and move to the modern movement in Chapter four.

Some aspects of skepticism can be traced back to thinkers more than two thousand years ago. As we will see, some of these ancient skeptics had something in common with modern skepticism, but in general their aims were rather different. Still it is worthwhile to look at the approaches of the ancient philosophers. Their ideas helped form the modern movement, and their thoughts are often fresh, vital and relevant today.

This is not a complete history of skepticism, and it does not mention all of the people who called themselves skeptics. The aim is simply to highlight some of the most important thinkers and exactly what they contributed to modern skepticism. We will start with a simple question: exactly what do we mean?

WHAT EXACTLY IS SKEPTICISM?

If we look at almost any definition of skepticism (or scepticism) in a dictionary, we will find one recurring element: the idea of doubt. Skeptics, we are told, doubt the truth of propositions. They may also doubt religious belief. Of course, as long as humanity has existed, people have made statements, and other people have doubted them. It is perfectly within the definition for skeptics to focus on the paranormal and to doubt the validity of paranormal claims to knowledge.

We can happily incorporate this into our understanding of skepticism: skeptics doubt the truth of paranormal claims; they are aware of human frailties and their concern is to arrive at the truth. As we will see, when we arrive at modern skepticism, this is how modern skeptics define their quest. First, though, we will look for the intellectual ancestors of modern skepticism. The key point is that we are looking for people who see doubt as part of the process of arriving at truth, and who are ultimately committed to the truth. A good place to start is with one of the greatest doubters of all, a man who died for his – thoroughly skeptical – beliefs: Socrates.

SOCRATES (~470 BCE–399 BCE)

Virtually all modern skeptics are happy to trace their thought back to Socrates. He was a shrewd thinker who would not stop asking difficult questions, who repeatedly reduced pompous know-alls to stuttering silence, and who lost his life for his skeptical principles. He was a formidable intellectual ancestor. At the same time, this is rather odd, to say nothing of non-skeptical, as the truth is that we know very little of what Socrates actually did or thought. Almost everything we know about this man comes from one source: the writings of the great philosopher Plato. Socrates himself wrote nothing. And we have very little corroboration of what Plato wrote about his teacher. The only other source of information that we have about Socrates is from the historian Xenophon, who presents us with a very different and far less interesting character. Essentially, Xenophon portrayed Socrates as a simple conservative, and the questioning, subversive aspects are not there.

It follows that we should take the personality and thought of Socrates with a large grain of skeptical salt. Plato was an important thinker, with controversial views on a range of subjects. It would be logical for him to give his own thought more credence by attributing it to his master, Socrates. Hence, we have no way of distinguishing what thought is Plato's and what is Socrates's. It makes intellectual life rather difficult (Taylor 1953).

Granting the problem, what picture of Socrates do we have? We know that he was a small, ugly man, who was apparently possessed of great physical strength and determination. His prime strategy was to engage fellow Athenians in conversation, and pretend that he knew nothing at all of the subject under discussion. Gradually, Socrates would draw from his victim contradictory assertions, and finally reduce him to stuttering incoherence with questions he was unable to answer.

What sort of awkward questions did Socrates ask? One of the best-known conversations was with an important and pompous citizen named Euthyphro (Plato 1953 [nd], pp. 303–40). This righteous man was on the way to prosecute his own father for alleged wrongdoing. Socrates engaged him in conversation, and soon managed to tie him in knots. For example, when Euthyphro proclaimed that he was doing what the gods wanted and what was just, Socrates asked him whether the gods wanted the actions because they were just, or whether they were just because the gods wanted them. It is a ferocious question, and any answer can be used to demolish Euthyphro's position. After more discussion Euthyphro, like many of Socrates's victims, discovered he had to be elsewhere quickly.

According to Plato's story, Socrates's subversiveness was so great that he was put on trial in Athens. The charge was: 1. of not worshipping the gods whom the state worships, but introducing new and unfamiliar religious practices and 2. of corrupting the young (Taylor 1953, p. 106).

Behind these vague words, apparently, lay the reality that Socrates was casting doubt on existing religious practices and also giving young people amusement by making their elders look stupid. These would not be crimes today. However, Athens was not a liberal democracy and Socrates was narrowly convicted by a jury of hundreds. His behaviour then was quite remarkable. He could have accepted banishment as an alternative punishment to death. Instead, Socrates proposed that he should receive a life pension (Plato 1953, pp. 341–66), and he was sentenced to death, taking poison at the age of seventy.

Many different philosophies and viewpoints have claimed Socrates as a patron. We can see that his questioning, doubting approach can mark him as a skeptic, but it is not clear exactly what he stood for. As far as we can tell, Socrates believed that no one sins wittingly (Plato 1953 [nd], pp. 287–8), so the more we know, the less likely we are to do wrong. He also did not dispute that we can know about the universe, his crusade seems to have been against pretentiousness and ignorance posing as knowledge. In sum, we do not know what Socrates really thought, but we can note the value of his questioning approach, as we will meet it again and again. In the sense of the definitions we are using, Socrates was a true skeptic. He doubted the truth of what was presented as knowledge, and he sought for knowledge himself.

PYRRHO OF ELIS (~360–~272 BCE)

From the vague birth and death dates of Pyrrho of Elis the reader can probably guess that we also know little about him. Like Socrates, Pyrrho wrote nothing, and we have to rely on the account of others about what he said, thought and did. The main secondary source, his disciple Timon's writings, are also incomplete. Pyrrho gave rise to an important school of skeptical thought, Pyrrhonism, which still has relevance today, as we will see. Of course, the same questions of reliability arise as with Socrates.

We do know that Pyrrho was a soldier in Alexander the Great's army, and that he gained some knowledge of the teachings of both Persian and Indian thinkers. His key point was that we cannot know anything objectively, and the true nature of the world must remain forever hidden from us (Russell 1971, p. 243). Timon, a successor to Pyrrho, took this even further, by arguing that equally good arguments can be found for and against any position.

Without going into the complexities of the arguments, it is clear that Pyrrho had a good point. We can observe objects and processes, but how can we know that what we observe has any relationship at all to what it actually is? The most we can ever say is that something's perceived properties consistently suggest some real properties, but we can never know for sure.

Of course, Pyrrho had no way of knowing that science would transform our view of the universe. His basic point is still perfectly correct, but we now have a body of knowledge which enables us to predict and manipulate the real world with a high degree of reliability. For example, we can never know for certain whether matter is really composed of atoms. All we can say is that the theory has been endlessly tested and, at the end of the day, no one has come up with anything better. It can then be argued that continuing to doubt in these circumstances is not a reasonable approach, it is simply playing an intellectual game.

What did Pyrrho deduce from his perspective? Since, he argued, nothing can be known, the person who is aware of this should not concern himself with unanswerable questions. There is no point seeking the unattainable. Instead, the wise person should cultivate a calm and equable manner, and live in accord with his society. Indeed, Pyrrho was noted for his almost superhuman calm and his indifference to the turmoil and disputes around him – although, apparently, he once lost this calm when he was attacked by a fierce dog.

Pyrrho's approach has a good deal in common with some modern trends of thought, such as postmodernism. Although the basis of the arguments differs – as does the aim behind it – the insistence on the failure of any viewpoint to establish itself as the truth is common to both. We should note, though, that the thoughts of both Socrates and Pyrrho had a feature that is mostly absent today. Both argued that adoption of their approach would help to make a person more virtuous. Indeed, the adoption of a virtuous lifestyle could reasonably be said to be the goal of both Socrates's and Pyrrho's philosophy. In this sense, their ultimate goal differed from that of the modern skeptics, who regard doubt as a necessary feature of the pursuit of knowledge.

Pyrrho's philosophy became one of the most coherent and powerful schools of Greek philosophy. The debates between Pyrrhonists and their main opponents, the Epicureans and the Stoics, were lengthy and sometimes acrimonious. However, there was another stream of skeptical thought which we might find more akin to modern skepticism. This went by the name of academic skepticism.

The first thing we should note is that the academic skeptics should not be confused with modern academics. The term 'academic' today refers to people working in academia. However, this is not where the ancient term 'academic skepticism' came from. The term was applied to a group of skeptics, more moderate than Pyrrho, who were associated with one of the academies, groups of scholars, which existed from time to time in Greece. Perhaps the two most important of these were Carneades and Aenesidemus.

Carneades (~214–129 BCE) had the same problem as Socrates: he displeased the government. However, the details are rather different. Carneades journeyed to Rome and gave a series of lectures in which he demonstrated that knowledge is impossible. He argued both sides of several questions, giving equally convincing arguments that opposed each other. The practical-minded Romans were unimpressed, and Carneades was banished (Russell 1971, pp. 245–6).

Aenesidemus (first century BCE) wrote ten arguments for skepticism. His arguments are usually taken as being the central theme of academic skepticism (Groarke 1990). The essential point is that while nothing can be known with certainty, dispassionate examination may show that some beliefs are more probable than others. In support of this argument, Aenesidemus pointed out that it is impossible to prove anything. Why? Because every assumption in a proof is based on some other assumption, or on an unproved assumption. Thus, we either find ourselves looking at unproved assumptions or an infinite regress of assumptions. Either way, nothing can be known with certainty. Although Aenesidemus is not a well-known figure in history, we will see his argument about uncertainty and probability recurring again and again. It is crucial to science and skepticism today.

Perhaps we should leave the arguments of the ancient skeptics there. We do not need to know very much. It is clear that their concerns were rather different from that of modern skeptics and, above all, the important terms 'science' and the 'paranormal' simply could not be part of their philosophy.

We can learn from their remarkable perceptiveness and their willingness to doubt everything, and to pursue chains of logic as far as they would go.

DOUBTING THOMAS AND PLINY THE ELDER

As the Roman Empire declined, ending with the sack of Rome in 410 (Gibbon 1879 [1782], p. 282), Christianity became more and more prominent. Finally, under Constantine, it became the official religion. The Roman Empire then fell into a long decline and, in western Europe at least, philosophy as the Greeks knew it no longer existed. For nearly a millennium western Europe was plunged into an age of faith, when the keenest minds – people like St Augustine and St Thomas Aquinas – were concerned with codifying, defending and justifying their religious views, rather than indulging in the kind of intellectual argument loved by the Greeks (Russell 1971). One can argue endlessly about the role of Christianity during this period. True questioning and doubt were discouraged and often brutally punished, but at the same time the Church did preserve a huge body of written material and learning which might otherwise have been lost. During this dark time the Arabic Islamic civilisation flourished, and later provided a stimulus for the rise of Europe.

In these long centuries we might look at just one skeptic, for his experience shows the position to which the doubter was reduced in this new age. We all know that Jesus died a hideous death on a cross. Christians also believe that he rose again from the dead, and was seen by many of the faithful. Naturally those who had seen Jesus ran to tell the others the joyful news. One of them, though, did not believe what he was told. This is the disciple Thomas who, the Bible tells us, reacted like this.

> He said 'Unless I see the mark of the nails on his hands, unless I put my finger into the place where the nails were, and my hand into his side, I will not believe it.'
>
> (*New English Bible* 1970, p. 187)

It took exactly a week for Thomas to receive the evidence he wanted, when this happened:

> Although the doors were locked, Jesus came and stood among them, saying, 'Peace be with you!' Then he said to Thomas, 'Reach your finger here; see my hands. Reach your hand here and put it into my side. Be unbelieving

no longer, but believe.' Thomas said, 'My Lord and my God!' Jesus said, 'Because you have seen me you have found faith. Happy are they who never saw me and yet have found faith.'

(*New English Bible* 1970, p. 187)

The term Doubting Thomas has often been used as a term of abuse, but consideration of this passage yields several interesting points. Thomas may have been a shrewd judge of human nature. The followers of Jesus must have been devastated by his crucifixion, and it is likely that many wild stories passed among them at this time. (For a modern version of this phenomenon, see Leon Festinger's classic book *When Prophecy Fails* (Festinger, Riecken & Schachter 1950), about a doomsday cult and the followers' reactions when doomsday did not come.)

So Thomas had perfectly good reasons for not believing wild tales about Jesus having appeared to other people. However, he did not ridicule the claims. Instead, he stated exactly what evidence he would require to be convinced of the resurrection. And a week later, Jesus came and gave him that evidence. Thomas behaved in immaculate skeptical fashion. Jesus did not condemn Thomas for his stance, nor did he rebuke him. He simply made the cryptic comment 'Happy are they who never saw me yet have found faith'. Thus, if modern skeptics are ever looking for a patron saint, they could adopt St Thomas! In the manner of a good modern skeptic, Thomas stated what evidence he would need in order to believe, and altered his view when the evidence was presented. Jesus's final comment shows the way that attitudes were moving: belief without evidence, provided it was the correct belief, of course, was beginning to be regarded as a virtue.

Thomas is a minor figure in the history of skepticism. There are others in this period. For example, there has been some discussion about whether Pliny the Elder (23–79 CE) could be classed as a skeptic. Jennifer Michael Hecht, in her monumental book *Doubt: A History* (Hecht 2003) thinks that he can. Parejko (2003) thinks that he can be regarded as a transitional figure, a man of his time but with tendencies towards rational doubt and inquiry. Having read Pliny's major work – or at least an abridged form of it (Pliny 1962 [nd]) – my conclusion is rather different.

Pliny is an impressive figure. He held the rank of a Roman Admiral, and died at Pompeii during the eruption of Vesuvius in that capacity. His *Natural History* is a huge compendium of what was believed at the time.

Reading the *Natural History* is both fascinating and exasperating. Pliny simply states what appear to be facts, sometimes expresses an opinion, and then moves on to something else. This passage on baldness might give some idea of what the style is like.

> Seldom do women shed their hair clean, and become bald: but never was there any gelded man known to be bald: nor any others that be pure virgins, and have not sacrificed unto *Venus*.
>
> (Pliny 1962 [nd], p. 131)

It is tempting to evaluate Pliny in terms of what we currently know. If we do this, then we conclude that the first two statements are correct: women seldom become bald, castrated men do not become bald. However, the next statement seems completely outlandish, as there is no evidence whatever that virginity has anything to do with male baldness. Still, that method of evaluating Pliny is unfair to him, and does not carry us far forward. It is easy to find absurd statements in which Pliny appears to believe. Here are two examples. The first is pointing to a cure for a heavy cold: 'If the rheum cause the mur, the pose or heaviness in the head, I find a pretty medicine to rid it away, by kissing only the little hairy muzzle of a mouse.' (Pliny 1962nd, p. 315) The second:

> And here I cannot choose but note unto you by the way, the strange property and wonderful nature that egg-shells have: for so hard compact and strong they be, that if you hold or set an egg headlong, no force nor weight whatever is able to break and crush it, so long as it stand straight and plumb upright, until such time as the head incline to a side and bend one way more than another.
>
> (Pliny 1962, p. 300)

One can only sympathise with the mouse in the first case, and marvel at the second statement. A small experiment – which I have carried out – shows its total falsity. Eggs are much stronger than one might think, but they are by no means indestructible. Why, one wonders, did Pliny not try the proposition out for himself?

There are some matters where Pliny plainly does not believe the stories. He denounces the story that semen will cure a scorpion sting (Parejko 2003: 42), but he believes that if a sting victim tells a donkey about it, the sting will be cured. He also disbelieves generally what magicians say (Pliny 1962 [nd], p. 314), even though they may sometimes have some show of truthfulness.

A useful question to ask is this: what criteria does Pliny use to tell the difference between truth and falsehood? He never really tells us. If Pliny had collected all fables together, then tried to sort out what was true and what was not, he would be one of the early and great skeptics. Of course, he would have been wrong some of the time, but all of us are. The key point regarding Pliny – and the reason why I cannot regard him as one of the precursors of modern skepticism – is not that he appears to be wrong a great deal of the time. It is that he makes no serious attempt to sort out what is right and what is wrong. Doubting Thomas spelt out what it would take to make him believe in Jesus's resurrection. Pliny does no such thing. As we shall see in the next chapter, one of Pliny's stories is an absolutely elegant example of Occam's razor – a key skeptical concept – but Pliny does not even hint at that important idea. For this reason, although Pliny has been very influential at times, and is still great fun to read, he must remain as a rather confused wanderer in a bewildering world, rather than one of its early cartographers.

Rather than spend more time on the ancient skeptics, it will be useful to move to a great transitional figure, a thinker who shows traces of the old, yet clearly took the first steps on the path to the new. This is the great French scientist, mathematician and philosopher René Descartes.

RENÉ DESCARTES: TO DOUBT EVERYTHING

René Descartes was a mass of contradictions. He was a devout Catholic, educated by the Jesuits, yet he set out to question absolutely everything. He had a comfortable private income and affected to be a gentleman of leisure, yet somehow he produced a vast mass of written work. He was a good scientist, an excellent mathematician and a truly great philosopher.

A good place to start is with the dates of Descartes's life, 1596–1650. He died before Newton published his great work, the *Principia,* in 1687, but he knew of the earlier scientists, such as Copernicus, Kepler and Galileo. Descartes had great insight into the implications of these developments. He realised that the argument over astronomy was not a matter of minor detail, but marked the rise of a new way of thinking. Descartes wanted to make use of that new way. He was not a great scientist, but he was an exceptionally clear thinker. So he set himself a quite remarkable project.

Descartes's goal was simply this. He would remain outwardly in conformity with his society – he lived in Holland and Germany, although he was a Frenchman. He would attend church, socialise with friends and pursue his normal life. Inside, though, he would set out to doubt everything. Nothing – absolutely nothing – would be accepted unless it could be proved. Descartes wrote that he could not prove that all his beliefs were false:

> But inasmuch as reason already persuades me that I ought no less carefully to withhold my assent from matters which are not entirely certain and indubitable than from those which appear to me manifestly to be false.
> (Descartes 1911 [1641], p. 9)

Descartes began from a position of total doubt, and would accept only that which could be shown with certainty. This is a bold approach to philosophy. However, by insisting on withholding consent 'from matters which are not entirely certain and indubitable', Descartes placed himself under a terrible handicap. This approach makes it impossible for Descartes to come to anything approaching a modern philosophy. It is remarkable how much progress he made.

Descartes was perfectly aware that merely seeing things did not prove them to be real. He knew that a large part of perception is bound up with mental processes. He was also aware of the possibility that the perceived world could be an illusion. He thought in terms of an 'arch deceiver' who could consistently manipulate what he saw, heard, touched and smelt so that he was systematically misled.

> I shall then suppose, not that God who is supremely good and the fountain of truth, but some evil genius not less powerful than deceitful, has employed his whole energies in deceiving me; I shall consider that the heavens, the earth, colors, figures, sound and all other external things are nought but the illusions and dreams of which this genius has availed himself in order to lay traps for my credulity; I shall consider myself as having no hand, no eyes, no flesh, no blood, nor any sense, yet falsely believing myself to possess all these things . . . and with firm purpose avoid giving credence to any false thing, or being imposed upon by this arch deceiver, however powerful and deceptive he may be.
> (Descartes 1911 [1641], p. 12)

Descartes' plunge into this abyss of doubt was traumatic, and was made much worse by the fact that he insisted on certainty before he would believe any proposition. Still, he found a way out. The first step he took is perhaps the most famous. If he thought, then clearly he had to exist. If he did not exist, then clearly he could not think. The phrase with which he stated this, *cogito ergo sum,* is rightly famous. One can sense the satisfaction in his writing as Descartes begins his climb out of the pit of absolute doubt.

But there is some deceiver or other, very powerful and very cunning, who ever employs his ingenuity in deceiving me. Then without doubt I exist also if he deceives me, and let him deceive me as much as he will, he can never cause me to be nothing so long as I think that I am something. So that after having reflected well and carefully examined all things, we must come to the definite conclusion that this proposition: I am, I exist, is necessarily true each time that I pronounce it, or that I mentally conceive it.

(Descartes 1911 [1641], p. 13)

Thus, Descartes existed, and could be certain of this fact. It would be absurd to talk about being deceived, or thinking and not existing. Modern philosophers have argued against Descartes' argument, but it is rightly celebrated and acknowledged (Russell 1971, p. 550). Still, having satisfied himself that he existed, what could he say about the world? This was much more difficult. Bearing in mind the possibility of the 'arch deceiver', how could Descartes get any further?

In hindsight, it is easy to criticise Descartes's next step. We could say that Descartes should have remembered Aenesidemus's approach; that although we cannot know for sure what is true, some theories are more likely than others. We cannot prove with certainty that the outside world exists, but we have never seen any evidence to contradict it. Thus, we can say with some assurance – though not total certainty – that there is an outside world resembling our perceptions.

However, this is not the route Descartes took. Instead, he introduced God. He used some rather unoriginal 'proofs' of the existence of God to show God existed. One is the famous ontological proof, devised by St Anselm. We can conceive of God, a being of whom none greater can be conceived. St Anselm went on to argue that if we say God does not exist then clearly God cannot be a being of which none greater can be conceived. After all, if we conceive this great God, and decide he does not exist, then we can conceive of an equally great God who does exist, and is thereby greater (Carey 2005).

This is an exasperating argument. Its refutation is simple in principle, but quite tricky in practice. Any of us can conceive of the greatest possible being. We can also imagine this being as existing or not. However, whether we *imagine* the being as existing or not tells us nothing about whether or not the being *actually exists*. As Paul Kurtz puts it:

The very fact that some individuals have an idea of God that entails existence as part of its meaning does not necessarily demonstrate that the idea must exist independent of the concepts in their minds.

(Kurtz 1992, p. 210)

We might also ask, rather pointedly, why it must be the Christian God which exists? Why not the God, or gods, of some other religion? Descartes did not address this, but regarded Anselm's argument, and a number of others, as conclusive proof that God existed. With God firmly in the picture, he then made a series of deductions about the world from them. For example, since God exists, He would not allow us to wander forever in error. There must be ways provided to correct error and find the truth. And since, by the most careful thinking and investigation, we can find no evidence that we are being deceived by a demon, we must conclude that the world exists. Descartes was out of his prison of doubt. However, in modern terms, his escape-hatch was not legitimate. Few of us today would accept 'proofs' of the existence of God, and then use that 'proven' existence to establish the existence of the world.

What might Descartes have done instead after his brilliant first step? He might have taken a leaf from the book of the earlier skeptics. As we have seen, they accepted that no propositions about the world could be shown with certainty, but they did argue that some propositions might be more probable than others. Thus, Descartes could have started with the certainty of his own existence, and then asked what propositions he could believe with great confidence, but not with absolute certainty. He might have noticed – indeed he did notice – that he always seemed to have two arms, two legs and so on, and so he could have concluded with great confidence that he actually had a body configured in this way. Then he might have noted that other people seemed to behave in ways which he could usually understand, and so could conclude – albeit with less confidence – that other people had thoughts, passions and fears very much like his own.

In this way, Descartes could have built himself a universe in which he could believe, though only one tiny part of it could be believed with complete certainty. However, Descartes did not choose this way. Although we would regard much of his thinking as flawed, there can be no doubt of his great intellectual courage, and the important contributions he made to modern thought.

DESCARTES'S CONTRIBUTION

What did Descartes contribute to modern thought? Essentially, he was the first thinker since ancient times to place doubt at the very heart of knowledge. Descartes saw clearly that without doubt there can be no knowledge, and he used this insight to construct a universe in which he could believe. The western world was beginning to emerge from an era in which unquestioning faith was considered a central virtue. Further-more, Descartes was a devout member of the major religious organisa-tion in western Europe, and so belief was, or could have been, imposed on him.

There is a whole philosophical industry that is focused on trying to work out what Descartes 'really' meant to do in his philosophical work. One of the most important and convincing arguments comes from the distin-guished thinker Marjorie Grene. Grene (1991) points out that Descartes differed from the ancient skeptics in a number of important respects. For example, the ancient skeptics did not doubt absolutely everything. They simply doubted that we can tell anything about the real nature of the world from appearances. Having established this, their point was that one should live a virtuous, equable life without worrying too much about the true nature of reality.

Descartes, by contrast, wanted to go far beyond this. His method of 'hyperbolic doubt' meant that appearances – and not just the reality behind them – came under question. Clearly, in our daily lives, we cannot seriously doubt everything. I am certainly not going to doubt whether I need to breathe, or whether eating is necessary to keep me alive. So what was Descartes doing? Grene thinks he was trying to do two things at once. One was to establish a logical foundation for a science of mathematical physics. The other was to produce a philosophy which was compatible with the doctrines of the Roman Catholic Church.

From this viewpoint, Descartes was using the tools of the skeptics for quite a distinct purpose. He was not concerned with a virtuous life – though he appears to have lived one – but used the weapon of doubt to clear away confusions about the beliefs which should, in his view, be founded on the rock of certainty. One of these beliefs was natural science, the other Roman Catholicism. Whatever our views, there is no doubt that Descartes's arguments set the course for a long argument to come. We can accept his courage in seeking to doubt, and his bold use of doubt to seek for knowledge, without necessarily accepting his other goals.

Another part of Descartes's thought should also be noted, although it does not relate directly to his skeptical work. It will become important later on in this book. Descartes did not merely argue that we should think in certain ways because it is more useful in finding the truth. He also argued that we have an ethical duty to think in certain ways. Indeed, Descartes can be argued to be the modern founder of the 'ethics of belief' or evidentialist movement (Bagger 2002). We will return to this theme later.

Descartes also seems to have spurred into action one of the few of his contemporaries, another Frenchman, Blaise Pascal (1623–1662), who was his intellectual match both as a philosopher and as a mathematician. In some ways Pascal's thought resembles Descartes, but Pascal never hid his strong commitment to his Catholic faith. Indeed, it has been suggested (Oakes 1999) that Descartes was the first modern philosopher, while Pascal can be regarded as the first modern Christian. This is because he acknowledged the doubt of Descartes and others, and yet set out to construct a philosophy which would render Christian belief rational. In fact Pascal went further than this, arguing that Christianity was the only true faith. His most notorious intellectual weapon was 'Pascal's Wager', an argument which suggested that non-believers should believe in God for their own self-interest. Perhaps his most powerful contribution, though, was his characterisation of humanity as a 'thinking reed'. That is, we are creatures who are physically frail and mortal and subject to a whole range of threats and dangers, and yet whose thought can encompass the cosmos. Although Pascal's work does not fall within the scope of this book, and indeed his thought is in many ways hostile to the skeptical tradition, he was an important and perceptive commentator on the human condition.

HUME ON MIRACLES

David Hume (1711–1765) was perhaps the greatest English-speaking philosopher of all time. He was not an Englishman, he was a Scotsman, and his pride in his nationality often comes through clearly in his work. Hume's arguments, and general skeptical approach, continue to be important to this day. Indeed, in the view of many modern philosophers, such as Russell (1971, p. 634), the problems he posed have never been fully answered. There is a journal, *Humean Studies*, which is devoted entirely to considerations of his work.

Despite this long-term influence, Hume was not a success in his own lifetime. His most substantial work, the *Treatise on Human Nature* (Hume 1986 [1739–40]) 'fell dead from the presses'. The great reactions of outrage and protest he anticipated never appeared. Hume's philosophy was largely ignored for much of his lifetime, and he never succeeded in being appointed to a professorship. Instead, he held a series of minor jobs, often of a non-academic nature. He worked as secretary to a general, and tutor to a madman. Indeed, he only managed to make himself financially secure when he wrote a best-selling history of England (Hume 1762)! Despite this, his influence is still with us today. Hume's key arguments can still be read, and are fresh and powerful. He is, in the intellectual sense, a living thinker and not a historical curiosity.

Hume aroused great controversy, with many people fiercely denouncing his arguments. However, almost all of them paid tribute to his kindly, agreeable nature, and it seems likely that the destructiveness of his arguments was to some extent mitigated by his own amiable personality.

HUME'S MAJOR THESIS

Hume's argument is put with great power and sophistication, but it centres on one simple point. In science, and in all other aspects of life, we are always concerned to ascribe causes to phenomena. If someone dies a strange death, we want to know what caused that death. If our car refuses to start, we want to know what is wrong with it. If the economy is going through a downturn, commentators ask why.

However, Hume's argument is that there is no way that we can ascribe causes with any certainty. Why not? Because all we can see are phenomena. If we want to say that A causes B, what evidence can we have for such a statement? All that we can say is that every time we have observed A, B has followed soon after. But, Hume points out, there is absolutely no reason to say that A causes B. We can only say that in every case we have seen, B follows.

It might be thought that we can improve on this. What if we know some of the mechanism which, we think, links A to B? Can we not then show that A causes B? No, because Hume will argue that we simply have a number of separate phenomena which we have observed to follow each other, and we cannot say that any of them causes the next.

Logically, Hume has destroyed large parts of the way we think about the world. Assuming the argument to be correct – and no-one has shown it to be wrong – then we cannot say anything with certainty about the way the world works, except that which we directly observe.

It should be clear that Hume's work is diametrically opposed to that of Descartes, although both can be classed as skeptics. Descartes, as we have seen, doubts all appearances but argues that there are certainties which can be established. Hume does not doubt appearances, but argues that we can infer nothing from them with any certainty. Rather than descending into this philosophical morass, we might simply note that Hume and Descartes, while apparently doubting everything around them, both got on with the job of living their lives and trying to affect the world for the better. And, as far as one can judge, they both made great contributions in that direction.

MIRACLES

Hume's contributions to philosophy are extremely important. For skeptics, though, his major work is elsewhere. It is a small essay – just over 20 pages. In this essay, 'Of Miracles,' (Hume 1980 [1748]), the great skeptic lays the intellectual foundation of modern skepticism. It is worth spending a little time on the argument, though the modern reader will find the original perfectly accessible: Hume was very good at saying exactly what he meant, and communicates well across two centuries.

'Of Miracles' was omitted from Hume's first book. It appeared later, and a great controversy exploded over the arguments.

Hume is very aware of the importance of what he is writing. Early in the essay, he introduces his argument as follows:

> I flatter myself, that I have discovered an argument... which, if just, will, with the wise and learned, be an everlasting check to all kinds of superstitious delusion, and consequently, will be useful as long as the world endures.
> (Hume 1980 [1748], p. 89)

After this immodest opening comes the argument. Hume's first point is that all of our knowledge comes from experience, and that this is not an infallible guide. Experience can lead us to make mistakes. Therefore, he reasons, using a famous phrase 'A wise man, therefore, proportions his belief to the evidence'. If the evidence is doubtful, and only weakly indicates a conclusion, then the 'wise man' will be very tentative in his judgements. If the evidence is overwhelming then, argues Hume, this can 'reasonably beget a pretty strong degree of assurance' (Hume 1980 [1748], p. 90). Where there is strong evidence on both sides, the assurance should be proportional to what is left after the subtraction of one set of evidence from the other.

Hume then moves on to miracles. His definition of a miracle is buried in a lengthy footnote. Though different in form, it resembles the definitions of the paranormal we looked at earlier: a miracle may be defined as, 'a transgression of a law of nature by a particular volition of the Deity, or by the interposition of some invisible agent' (Hume 1980 [1748], p. 93).

Note that Hume's definition is similar to those of the paranormal: it is an inexplicable observation or event which transgresses a law of nature. His definition does go rather beyond the earlier definitions, as it also includes acts of the deity. Since we are not considering religion as part of the paranormal, this need not worry us, and we will simply note that Hume's view of miracles strongly resembles our general view of the paranormal, in that miracles are both strange and inexplicable. But how does Hume put together this idea with his earlier observations about wise people and the evidence? This is the heart of his argument. Hume stresses that he is focusing on eyewitness testimony for miracles, though his argument can be applied to all types of evidence. Then he turns to miracles in this way:

> Suppose, for instance, that the fact, which the testimony endeavours to establish, partakes of the extraordinary and the marvellous; in that case, the evidence . . . admits of a diminution, greater or less, in proportion as the fact is more or less unusual.
>
> (Hume 1980 [1748], p. 91)

By definition, a miracle is an astonishing event. If it were not astonishing, if it did not transgress the laws of nature, it would not be a miracle. Therefore any miraculous event has arrayed against it the whole of human experience, and the evidence can never be strong enough for a 'wise man' to conclude that it has been established. Then Hume states his 'general maxim':

> That no testimony is sufficient to establish a miracle, unless the testimony be of such a kind, that its falsehood would be more miraculous, than the fact, which it endeavours to establish; and even in that case there is a mutual destruction of arguments, and the superior only gives us an assurance suitable to the degree of force, which remains, after deducting the inferior.'
>
> (Hume 1980 [1748], p. 94)

Then Hume gives an example of how his principle works. If someone tells him that he saw a dead man restored to life, how does Hume think about it?

> I immediately consider with myself, whether it be more probable, that this person should either deceive or be deceived, or that the fact which he relates, should really have happened . . . and always reject the greater miracle. If the

falsehood of his testimony would be more miraculous, than the event which he relates; then, and not till then, can he pretend to command my belief or option.

(Hume 1980 [1748], p. 94)

Hume's argument is simply that, from the very definition of miracles (and, incidentally, from the nature of the paranormal generally) there can never be enough evidence for a wise person to believe the claims. The argument is profoundly important – as important as Hume thought it was. It uses the basic nature of the paranormal to exclude the possibility of rational belief in paranormal claims.

Having outlined his major argument, Hume makes a number of other points. Although not as powerful, they are also important and worth noting. First, he points out that in all history there is no case of testimony to a miracle made by enough people – beyond suspicion of deceit or error – as to assure us of the truth of their testimony. Second, he points out that humans seem to be strongly tempted to believe in the paranormal: 'The passion of surprise and wonder, arising from miracles, being an agreeable emotion, gives a sensible tendency towards the belief of those events, from which it is derived' (Hume 1980 [1748], p. 95). So we all are tempted to believe in the paranormal. We noted this point in Chapter one, where it constitutes one of the reasons why we tend to believe in the paranormal.

Hume's third point may look a little strange to us in the twenty-first century. He points out that miracles seem 'chiefly to abound among ignorant and barbarous nations'. This is rather out of date, considering the way that the New Age and related philosophies have blossomed in California, one of the most advanced and sophisticated parts of this planet. Finally, Hume points out that the miracles which support one faith are opposed by the claims of miracles which are claimed for another faith. In consequence, the evidence for one religion's miracles is cancelled out by the evidence for another religion's miracles, weakening their credibility yet further.

These later points are worth making today, except, perhaps, for the third, but it is the first argument, the one that gave rise to Hume's general maxim, which is the most important. If this argument is justified then, as Hume says, it is a powerful argument against believing any claims of miracles, and of any paranormal phenomena, no matter how much evidence is produced in their favour. But, is Hume's argument justified? It is worth making at least a few simple points on this question.

Hume's argument rests on two premises. One is that a wise person proportions belief to the evidence – or at least to the balance of the evidence. The second premise is that we know a good deal about how the world actually works, from the accumulated understandings of humanity. Balanced against the latter, any claim is going to look ill-supported. If we accept Hume's argument in its entirety, how can we accept any major change in our worldview at all? As Arthur C. Clarke (1962, p. 27) argues, new technology can often be quite beyond belief, and indeed can look like magic. For example, if we bang a couple of pieces of metal together in the right way, we obtain an explosion which can destroy a city. Surely, this is a 'miracle' as profound as any described by Hume: it is completely outside our experience and contrary to it. Therefore, why should any of us believe in atomic bombs? Of course, there is a mass of evidence; thousands of people experienced the atomic bombs at Hiroshima and Nagasaki, or saw the consequences of them (my father was one such witness). Yet as Hume would point out, this evidence is slight compared to the mass of knowledge about the way the world normally is.

Later in his famous paper, after pronouncing his 'general maxim' and illustrating its devastating effects, Hume seems to be saying something more moderate. He imagines the case where evidence from across the world agreed on the occurrence of a remarkable event.

> Thus, suppose, all authors, in all languages, agree, that from the first of January 1600, there was a total darkness over the whole earth for eight days; suppose that the tradition of this extraordinary event is still strong and lively among the people: that all travellers, who return from foreign countries, bring us accounts of the same tradition, without the least variation or contradiction: it is evident, that our present philosophers, instead of doubting the fact, ought to receive it as certain, and ought to search for the causes whence it might be derived.
>
> (Hume 1980 [1748], pp. 105–6)

What is Hume saying? At the end, he pronounces that the days of darkness might be believable if the testimony for it is 'very extensive and uniform'. However, the point of his general maxim is that no matter how much evidence may be produced – and no matter how uniform it may be – it still cannot override the immense mass of evidence pressing against it. So why has Hume changed his position? There seem to be two reasons. One is that this imaginary account of darkness is not being used to found a religion, the other is that the darkness is consistent with the 'decay, corruption and dissolution of nature'.

These reasons are not very satisfactory. It might be worth noting that parts of the world are in continuous darkness for months at a time, during the Arctic and Antarctic winters. Thus, extended darkness is actually commonplace on some parts of the Earth. Perhaps Hume is influenced by the relative lack of miraculousness in this case.

However, things then get worse. Hume gives yet another case of a claimed miracle, which he would not believe at all. This concerns Queen Elizabeth the First, and runs like this:

> But suppose, that all the historians who treat of England, should agree that, on the first of January 1600, Queen Elizabeth died; that both before and after her death she was seen by her physicians and the whole court, as is usual with persons of her rank; that her successor was acknowledged and proclaimed by the parliament; and that, after being interred a month, she again appeared, resumed the throne and governed England for three more years.
>
> (Hume 1980 [1748], p. 106)

Hume accepts that the darkness in 1600, if properly supported by evidence, is credible, but apparently nothing will persuade him that Elizabeth the First could live after dying. He gives his reasons this way:

> ... the knavery and folly of men are such common phenomena, that I should rather believe the most extraordinary events to arise from their concurrence, than admit of so signal a violation of the laws of nature.
>
> (Hume 1980 [1748], p. 106)

Let us agree to Hume's point about the folly and knavery of people, and that if Elizabeth I were supposed to have risen from the dead, it would far more likely be some sort of deception rather than a genuine event. On the other hand, this 'miracle' has nothing to do with religion: Elizabeth was a monarch, not a prophet. And it is not at all clear why, if enough people testify that Elizabeth did die, Hume would find this story unacceptable, but swallow the claims about the days of darkness.

We have to conclude that Hume is not completely consistent. The major part of his essay is devoted to establishing that there can never be enough evidence for miracles, then he proposes a miracle – or at least an extremely unusual event – for which there might be enough evidence, then he proposes another miracle for which there cannot be enough evidence, and it is all rather unclear.

The problem here is a genuine one. Hume has to reconcile a principled argument that there can never be enough evidence with the pragmatic fact that evidence can establish almost any proposition (Burns 1981, p. 154). How do we resolve this paradox? One way is to be a little less fierce than Hume. Perhaps our knowledge of the way the world works is always too limited. Given plausible testimony, from several different sources, perhaps our attitude towards miracles might be that although they should be viewed with skepticism, there are certainly more strange things in heaven and earth than are dreamed of in our day-to-day lives. In this way, without succumbing to every miracle-monger, we might at least be able to avoid rejecting everything that is strange and unprecedented. As we will see, modern skeptics take a less drastic position than that of Hume.

HUME'S CONTRIBUTIONS TO MODERN SKEPTICAL THOUGHT

We will not attempt to evaluate the whole of Hume's thought. However, we can look at some of his contributions to skeptical thought, and it is clear that here he is one of the great pioneers.

First, and it is easy to forget this, Hume has highlighted the intrinsic limitations of our own knowledge. We are creatures in the world, and we should never give in to the temptation of assuming that we know everything, or that we can know everything.

Second, Hume has shown that we are not faced with a 'yes or no' decision about belief or unbelief on any topic. We can believe tentatively, or with a reasonable degree of assurance, while still bearing in mind our uncertainty. In this, Hume went beyond Descartes, who insisted on certainty.

Third, Hume stressed that some beliefs, by their very nature, require more evidence than others – and beliefs in miracles are the hardest of all to accept. Whatever the belief, we must weigh the evidence for and against the belief. We must also have some idea of what strength of evidence is necessary to accept the belief it, and then we can come to a reasoned conclusion.

It is always easy to judge with hindsight, but it is valuable to look at the skepticism of the Greeks, of Descartes and of Hume from today's perspective. Clearly, one enormous difference between the Greeks and the others is that they had no sense of the steady, cumulative ability of science to produce reliable explanations of the workings of the universe. This process had barely begun during their lifetimes. Their lack of this knowledge seems to have made their entire philosophy deficient from our viewpoint, as they struggled to make sense of what they saw.

By contrast, both Descartes and Hume were aware of the explanatory power of science, and both built it into their philosophies. From our perspective, Descartes bravely set out to doubt everything, but found an escape-hatch which enabled him to avoid living with this doubt. Hume, on the other hand, readily accepted doubt and sought to formulate a philosophy which would enable him to live with it. In doing so, he laid the foundations of modern skepticism, which we will examine in Chapter four.

LAUNCHING PAD

Russell's majestic *History of Western Philosophy* (Russell 1971), written in a clear style, is an indispensable resource for a picture of the broad sweep of Western philosophy. Russell has his biases, and simply cannot see the value of some philosophers, but his writing is so clear that the non-philosopher gains insight into a great deal of thought. He has a section on the French encyclopaedist movement, which is worth a look.

Of course, David Hume's (1980 [1748]) seminal skeptical work is important, and Hume can also make his thoughts perfectly clear, despite the passage of two centuries. The only irritation to the modern reader must be that Hume has the habit of scattering commas, apparently at random, throughout his work.

If it is possible to obtain, Paul Kurtz's book *The New Skepticism* (Kurtz 1992) is worth reading, as it gives the views of perhaps one of the most powerful modern thinkers on his forbears.

For more information on almost any of the major skeptical thinkers, the *Stanford Encyclopedia of Philosophy* (2006), an online document, is a wonderful place to start. It has extensive articles on almost all facets of the discipline, usually written in an accessible style.

4 | Modern skepticism

W HERE SHALL WE set the beginning of modern skepticism? As we have seen, the modern skeptical movement can trace its roots back to David Hume and René Descartes in the Enlightenment. Francis Bacon, a courtier in the court of King Henry VIII, could be called a skeptic. We could also look at many important skeptics in the nineteenth century, such as Herbert Spencer, and early twentieth century skeptics such as Harry Houdini (1980). Later on we will look at one of these early skeptics, William Clifford. However, most modern skeptics regard a clear and simple sequence of events as marking the founding of the modern movement.

The key event is the publication in 1952 of a book by the polymath writer Martin Gardner, titled *In the Name of Science*. The book did not sell well in its original incarnation, but it was reissued in 1957 under a different title, *Fads and Fallacies in the Name of Science* (Gardner 1957) and it has remained in print ever since. Half a century later, its contents are intriguing, and we will review them in a moment.

By itself, Gardner's book made little perceptible difference. However, in 1975 a statement appeared in *Science News* titled 'Objections to Astrology' (Bok et al 1975), signed by a number of Nobel Prize winning scientists. The following year a conference was held in Buffalo, New York, on 'The New Irrationalisms: Antiscience and Pseudoscience'. From this came the Committee for the Scientific Investigation of Claims of the Paranormal (CSICOP) and their magazine, eventually titled the *Skeptical Inquirer*. Groups under the aegis of CSICOP appeared across the US, and the skeptical movement also took root in other countries. One of the most successful of these was the Australian Skeptics, with their magazine the *Skeptic*.

What did these early modern skeptics have to say? Were their concerns similar to those of today, or were they different? There is no doubt that the founding skeptics in the 1970s were concerned about the onrush of mystical and pseudoscientific ideas which emerged in the late sixties and early seventies. Even before that, in the 1950s, it is clear that the early skeptics were, at least partly, reacting to different paranormal claims from those of today. A glance down the contents page of Gardner's book tells a fascinating story. Some of the chapter headings could belong to a book published today. Consider, for example 'Geology versus Genesis', 'ESP and PK', 'Flying Saucers' and 'Medical Cults'. Almost any survey of twenty-first century paranormal phenomena would have to include these items, though some might have different names. Flying saucers, for example, are now known as UFOs, and medical cults are now called alternative and complementary medicines. The content is often similar. By contrast, some of the chapters seem to come from another world. Who today is interested in 'Throw Away Your Glasses!', 'Sir Isaac Babson', 'Zig-zag and Swirl' and 'General Semantics'? Most modern readers will have no idea what these are about.

These changes suggest that some aspects of paranormality have 'legs'; they persist decade after decade, building on old claims and adding new ones. By contrast, other claims fade into obscurity, perhaps with the death of their most prominent advocates. It would be interesting to know why some claims persist and others fade, but so far little research has been done on this topic.

There is another feature of Gardner's book. The book does not actually descend into abuse, but is characterised by a rather harsh 'debunking' tone. Gardner himself is, by all accounts, a kind and thoughtful man. However, the aim of the book appears to be to 'debunk the punks'. Here is an example. In Gardner's view, the pseudoscientist exhibits these traits:

1. He considers himself a genius.
2. He regards his colleagues, without exception, as ignorant blockheads . . . Frequently he insults his opponents by accusing them of stupidity, dishonesty, or other base motives . . .
3. He believes himself unjustly persecuted and discriminated against . . . It is all part of a dastardly plot . . .
4. He has strong compulsions to focus his attacks on the greatest scientists and the best-established theories. When Newton was the outstanding name in physics, eccentric works in that science were violently anti-Newton. Today, with Einstein the father-symbol of authority,

a crank theory of physics is likely to attack Einstein in the name of Newton...

5. He often has a tendency to write in a complex jargon, in many cases making use of terms and phrases he himself has coined... Many of the classics of crackpot science exhibit a neologistic tendency.

(Gardner 1957, pp. 13–4)

Much of this is perceptive: it is astonishing how many paranormalists do conform to Gardner's picture. However, the severity of the language adds to this. It is possible that a gentler tone might have had more influence. The skeptical movement is still divided today between those whose goal is to 'debunk' the paranormalists, and those whose goal is to investigate their claims. Of course, the two overlap to a great extent, but the tension is quite noticeable in the modern skeptical movement (Clark 1997, pp. 125–6).

Gardner also foreshadows one of the major reasons for the appearance of modern skepticism. This is the feeling that the modern pseudoscientist is a parasite on the – genuine and important – work of modern science. In the introduction he writes of the pseudoscientist:

> He is riding into prominence, so to speak, on the coat-tails of reputable investigators. The scientists themselves, of course, pay very little attention to him. They are too busy with more important matters. But the less informed general public, hungry for sensational discoveries and quick panaceas, often provides him with a noisy and enthusiastic following.
>
> (Gardner 1957, p. 3)

This theme, that paranormal claims may be replacing science in the esteem of the public, appears again and again in the writings of the founders of the modern skeptical movement. Consider for example, Paul Kurtz, who founded CSICOP. He has perhaps the most powerful intellect in the modern skeptical movement, and describes his motivations in these terms:

> I was distressed that my students confused astrology with astronomy, accepted pyramid power, the Loch Ness monster, Kirlian photography, and psychic surgery without the benefit of a scientific critique. Most of my scientific colleagues were equally perplexed by what was happening, but they were focused on their own narrow specialities – interdisciplinary efforts were frowned upon – and they did not know what the facts of the case were.
>
> (Kurtz 2001b, p. 42)

Kurtz agrees with Gardner, arguing that the scientists are doing their own work, and less educated people cannot tell the difference between scientific and paranormal claims. Therefore, something has to be done.

Kendrick Frazier, who has edited the *Skeptical Inquirer* since 1977, makes a similar point when he talks about opinion polls:

> They also show that the public doesn't know much about science. This is especially true of the methods and processes of science. Some understanding of these processes is really necessary for getting a sense of what is good science and what is bad science (or worse).
>
> (Frazier 2001, p. 49)

The pictures painted by Gardner, Kurtz and Frazier are consistent. Science is, they argue, an enormously valuable way of finding out about the universe, but the public has no means, at present, of discriminating between genuine science and the claims of the paranormalists. The scientists, who can tell the difference and could educate the public, are involved in their research. This gives modern skepticism its role. Modern skepticism can investigate the paranormal and can educate the public in telling the difference between real science and the paranormal.

These are praiseworthy concerns. It is important that the general public, or as many of them as possible, should be able to distinguish between the real and the bogus. Otherwise it is hard to see how a modern society, let alone a democratic political system, can hope to survive. The concerns of Gardner, Kurtz and Frazier form a foundation for the modern skeptical movement.

In this chapter, we will take a similar approach to the chapters on science and the paranormal. We will look at the meaning of modern skepticism, then go on to examine its place in society.

THE GOALS OF THE SKEPTICS

What are the goals of the modern skeptical movement? What a movement actually does, of course, may be different from its formal goals. Still, the goals represent ideals to which skeptics aspire, and skeptical movements are remarkably consistent in what they set out to do. For example, if one visits CSICOP's website (now renamed CSI), the goals are easy to find. The aims of CSICOP are as follows (we shall see later that they have recently been changed):

The Committee for the Scientific Investigation of Claims of the Paranormal encourages the critical investigation of paranormal and fringe-science claims from a responsible, scientific point of view and disseminates factual information about the results of such inquiries to the scientific community and the public. It also promotes science and scientific inquiry, critical thinking, science education, and the use of reason in examining important issues.

(CSICOP 2004)

The main goal involves examining the paranormal from a 'responsible scientific point of view'. This, of course, reflects the concerns of the founding skeptics. Their concern was that the scientific case against the paranormal was going by default, and they set out to do something about it. However, as I shall argue a little later in this chapter, if they are to be effective, the skeptics cannot confine themselves to this particular activity. They must do a great deal more or they will fail. Before elaborating this point, let us examine another national skeptical organisation, perhaps the most successful one in the world, in terms of its population. The Australian Skeptics, founded in 1981, have similar goals to CSICOP, and these appear on their website as follows:

To investigate claims of pseudoscientific, paranormal and similarly anomalous phenomena from a responsible, scientific point of view. To accept explanations and hypotheses about paranormal occurrences only after good evidence has been adduced, which directly or indirectly supports such hypotheses.

To encourage Australians and the Australian news media to adopt a critical attitude towards paranormal claims and to understand that to introduce or to entertain a hypothesis does not constitute confirmation or proof of that hypothesis.

To publicise the results of these investigations and, where appropriate, to draw attention to the possibility of natural and ordinary explanations of such phenomena.

To stimulate inquiry and the quest for truth, wherever it leads.

(Australian Skeptics 2008)

In both cases, the skeptics are committed primarily to the responsible scientific investigation of claims of the paranormal. There is nothing wrong with this goal, except that it is inadequate. Skeptics do not simply carry out scientific investigations of the paranormal. What is more, if they did, they would be nowhere near as effective as they are. We will look at why this must be so, after examining the skeptic's intellectual tools and approaches.

INTELLECTUAL TOOLS OF THE MODERN SKEPTICS

What are the principles which guide the behaviour of modern skepticism? As we shall see, skepticism is a grassroots movement, and there is no skeptical Pope or source of authority. As a result, there are many skeptical approaches, and ideas. Therefore, we will look at some broad principles that seem to underlie skeptical investigation, and in subsequent chapters we will try to see how they work in practice.

There is no cookbook approach to skepticism any more than there is to science, although some principles are extremely important and recur again and again in the skeptical approach. For more details, the reader can turn to the excellent paper by Caso (2002). In this paper, Caso outlines what he describes as three debating tricks used by skeptics. After long reflection, I have come to the conclusion that they are much more than debating tricks: they are vital principles, and understanding them leads to a good grasp of what the skeptics are about. We will look at the three key principles – burden of proof, Occam's razor and Sagan's balance, and see why each fits well into the skeptical approach.

BURDEN OF PROOF OR BURDEN OF EVIDENCE

The first principle is the burden of proof. Many of us know about the burden of proof from watching a criminal trial on television or in the cinema. Someone is on trial, accused of committing a crime. The prosecution attempts to prove beyond reasonable doubt that the defendant is guilty of the crime and produces evidence to show this is so. The defence attempts to cast doubt on the prosecution's case, and perhaps also tries to show that the crime could have been committed by other people. The jury then decides whether the prosecution has proved its case beyond a reasonable doubt. In most countries, the jury must render a verdict of either 'guilty' or 'not guilty', though in Scotland the verdict of 'not proven' is also allowed.

There are other legal arrangements. In civil cases, for example, where people sue each other for damages, proof beyond a reasonable doubt is not required. In such cases the 'balance of probability' is used. Neither side need prove its case beyond a reasonable doubt, but only establish that its claims are more likely than the claims of the other side.

In most courtroom situations, a clear burden of proof is placed. One side or the other must prove its case, or at least show that the balance of evidence is in its favour. Does this resemble at all the situation where a paranormal claim is made? In the skeptics' view, there is a clear burden of proof upon the claimant. As Paul Kurtz says 'If someone were to claim that mermaids exist, the burden of proof is upon him, not the skeptic to disprove the fact.' (Kurtz 1998, p. 26)

Why should this be? Is it not unfair to burden the paranormalist with the task of establishing that the belief is justified? The answer has echoes of David Hume. We do not know everything about the universe – not by a long way – but we do know some things. The knowledge that we have is of value, and it should not lightly be disregarded. And, as we have already seen, there is an enormous mass of different, and often contradictory, paranormal claims in existence. If we accept the task of disproving them all, then not only do we disregard the knowledge we already have, we also place ourselves in an impossible position. How do we decide which of this mass of paranormal claims to disprove and, in the meantime, which do we decide to believe? On the other hand, if we place the burden of proof on the paranormalist, the knowledge we have about the universe remains as it is, until the evidence is strong enough to persuade us to modify it (Caso 2002). In short, placing the burden of proof anywhere but on the paranormalist leads to absurdities which cannot be reconciled. Placing it on the paranormalist leads to a logically consistent position and enables us to maintain our current beliefs until the evidence is strong enough for a change.

Of course, in general, the burden of 'proof' does not involve proof at all. We do not require ufologists to *prove* that UFOs are alien spacecraft, we simply want them to produce enough good quality evidence to take the claim seriously. The same applies to all aspects of the paranormal: proof is not needed, evidence is. For this reason we might call the burden of proof the burden of evidence, but the older name is more familiar.

Of course, this leaves unanswered the question of how much evidence is needed. The third principle – Sagan's balance – gives us some guidance in this direction, and we will know more after looking at a few cases in the next two chapters.

OCCAM'S RAZOR

The second skeptical tool is Occam's razor. This is named after William of Ockham (or Occam), a medieval churchman who devised one of the most useful skeptical tools in existence (Caso 2002). We do not know much about William: his place of birth and the spelling of his name are both in doubt (Russell 1971, pp. 459–65). Very simply, Occam's razor boils down to this short maxim: 'It is vain to do with more what can be done with fewer'.

There are assorted complexities associated with Occam's razor. After all, it was formulated by a strong religious believer, in a time well before the existence of modern science. However, as it relates to skepticism, its main point is simply this: if a natural explanation exists for a phenomenon, then there is no point in having a paranormal explanation as well.

Let's take an example. Imagine that I claim that I have psychokinetic powers. That is, by the power of my mind, I state that I can cause objects to move without any apparent force being applied. I demonstrate this to you. In my presence pencils roll on tables and playing cards seem to hang in the air. You are very impressed. Then you learn that I am an internationally known magician, who has spent years practising exactly these illusions. All of a sudden there is a perfectly ordinary explanation for the amazing events you have seen. I have been fooling you, and the need for a psychokinetic explanation disappears.

This is the heart of the value of Occam's razor to skeptics. It provides that wherever a natural explanation exists for a phenomenon, the need for a paranormal one disappears. Thus, skeptics can look for natural explanations for clairvoyance, faith healing or ESP knowing that if they find one, the need for the paranormal claim simply is eliminated. Occam's razor is a vital part of the modern skeptic's armoury, and essential to clear thinking generally. Of course, Occam's razor is also used in science, where there is a preference for simpler, more elegant theories over complex ones.

CRESINUS'S PROBLEM AND HOW OCCAM'S RAZOR SOLVED IT

While we are looking at Occam's razor, it is worth examining an example from a bygone era. The chronicler of the case did not recognise the principle involved, though to a skeptic it is likely to stand out clearly.

We have already looked at the amazing work of Pliny the Elder. He cannot properly be called a skeptic, as it is not clear on what basis he tries to sort out truth from falsehood. In his rambling *The History of the World*, Pliny (1962[n.d.]) gives an excellent example of Occam's razor in action, though he shows no sign of seeing the logic which underlies it. Pliny's story begins with a lowly farmer who is rather too good at his work:

> There was one *C. Furius Cresinus*, late a bondslave, and newly enfranchised,
> who after that he was set at liberty, purchased a very little piece of ground,
> out of which he gathered much more commodity than all his neighbours
> about him out of their great and large possessions: whereupon he grew to
> be greatly envied and hated, insomuch, as they charged him with indirect
> means, as if he had used sorcery, and by charms and witchcraft drawn into
> his own ground that increase of fruits which should otherwise have grown
> in his neighbours' fields.
>
> (Pliny 1962 [nd], p. x)

The picture is a vivid one. The lowly ex-slave outproduces the farmers
around him, and the resentful muttering starts. (One can imagine the
furtive discussions: 'It cannot possibly be that *we* are not doing as well as
we could. *He* must be doing something wrong. I bet it's sorcery.')

Cresinus was indicted. This was not a time of hysterical witch-hunts,
but he knew that if he were found guilty of using sorcery, charms and
witchcraft, he would face a large fine. The problem was, what could he do?
How do you prove that you are not guilty of doing these illegal activities?
In later ages, this impossible problem led to the death of tens of thousand
of accused 'witches' (e.g. Trevor-Roper 1969; Shermer 1997). Cresinus was
almost certainly uneducated, but I suspect he was a man of considerable
ability. When the day of the trial came, he had his defence ready. As Pliny
tells us he presented to the court his strong, hard-working daughter, 'well
fed and well clad . . . his tools and plough irons of the best making, and
kept in as good order; main and heavy coulters, strong and tough spades,
massy and weighty plough-shares, and withal his draught oxen, full and
fair' (Pliny 1962, p. 167).

This was a good start to his defence. Cresinus was showing that he
had good equipment, good farm animals and a strong and hard-working
daughter. What is more, he took good care of all of them. Then he began
to speak.

> 'My masters' quoth he 'you that are citizens of Rome, behold these are
> the sorceries, charms, and all the enchantments that I use' (pointing to his
> daughter, his oxen, and furniture abovenamed;) 'I might besides' (quoth he)
> 'allege mine own travail and toil that I take, the early rising and late sitting
> up so ordinary with me, the careful watching that I usually abide, and the
> painful sweats which I daily endure; but I am not able to represent these to
> your view, nor to bring them hither with me into this assembly.'
>
> (Pliny 1962, pp. 167–8)

This is a clever speech. Cresinus begins with a joke. He points to the humdrum means by which he raises his crops – the spades, ploughs, oxen and so on – and refers to them as his sorceries, charms and enchantments. Then comes a master stroke. He diverts attention from the fact that he cannot prove his innocence, and instead talks about something equally important. He points to the daily slog of work on his farm, the hours he puts in and the sweat he produces. Now perhaps many of his hearers would have seen Cresinus or his daughter working long hours on the property. Although he cannot 'bring hither' this evidence, his hearers might well have known how hard he worked, and have accepted this point as well.

How did the assembled citizens react to Cresinus's defence? Pliny tells us the verdict was swift:

> The people no sooner heard this plea of his, but with one voice they all acquit him and declared him unguilty without any contradiction.
>
> (Pliny 1962 [nd], p. 168)

Pliny's conclusions from this story are rather disappointing. He hardly goes beyond what Cresinus himself said.

> By which example, verily, a man may soon see, that good husbandry goeth not all by much expense: but it is painstaking and careful diligence that doth the deed.
>
> (Pliny 1962 [nd], p. 168)

In fact, Cresinus was using a form of Occam's razor in his argument. Faced with a charge of using magic to increase his crops, he showed that he had good equipment and maintained it well, that he had good help on his farm, and he probably appealed to common knowledge in saying that he was a hard worker. In short, Cresinus was arguing that his hearers *did not need* magic to explain how he raised good crops: there were perfectly good explanations without resorting to witchcraft. And, since Cresinus was able to explain his prosperity in mundane terms, there was no point in multiplying the explanatory entities. It is hard to think of a more elegant example of Occam's razor – more than a millennium *before* William of Occam.

SAGAN'S BALANCE, WITH A BOW TO DAVID HUME

The third key tool in the skeptic's armoury is what Caso (2002) refers to as Sagan's balance. One could equally easily refer to it as Hume's balance, for its roots in that great thinker's argument are perfectly clear. Still, Hume's place in the development of skepticism is completely secure, and there is no harm in recognising Sagan as well. After all, Sagan was not only a distinguished scientist and populariser of science, he was also an outspoken skeptic, ranking in the top ten skeptics of the twentieth century (Anonymous 2000). Essentially, Sagan's balance postulates that extraordinary claims require extraordinary evidence, a direct analogy to Hume's argument that the wise person proportions belief to the evidence.

How does Sagan's balance work? Here is an example. My workplace is on Griffith University's Nathan campus. Running around that campus is a ring road, three kilometres of bitumen upon which cars are almost always to be seen. Therefore, if someone comes to me and announces that they have seen a car on the ring road, my only surprise is that something so trivial would be mentioned. It would be a surprising day to find no cars on the Griffith University ring road. The claim is ordinary, and little or no evidence is required to establish it in my mind.

Now imagine that I am told that the Prime Minister has been seen in a car on the ring road. This is a little unusual, though Prime Ministers have visited the campus before. It is a somewhat extraordinary claim, and I would require extra evidence to accept it. Perhaps a newspaper has a picture of the Prime Minister at the University, or several people, quite independently, tell me they have seen the Prime Minister on campus. The claim is not all that astonishing, so only a modest amount of evidence is needed: I can be convinced without too much trouble.

Now imagine I am told that a pink translucent UFO has been seen on the ring road, with a dinosaur looking out of a porthole. This is a completely extraordinary claim, and the evidence required would have to be very strong. Indeed, David Hume would argue that the evidence would have to be so strong that it can overturn all of our existing systems of belief. Even if we do not take that position, we all know that eyewitnesses can be misled, or lie. We know that photographs can be faked, as can film, and that even our own perceptions can be mistaken. Thus, it would take a very great deal of evidence to convince a reasonable person that they saw the UFO. Any lesser amount would be weighed in Sagan's balance and found wanting.

Sagan's balance is especially valuable because it gives us a guide to how much evidence would be needed to accept a claim. If someone has an astonishing claim, but only a minor piece of evidence, then it is weighed in Sagan's balance and found wanting. Another situation is when an amazing paranormal claim is made, and evidence is produced which might be enough for Sagan's balance to be tipped in the direction of belief. However, careful examination then shows that the evidence is not as strong as the claimants thought. Then, Sagan's balance is no longer tipped and the claim should not be believed. We will see a case of this in the next section.

These three tools – burden of proof, Occam's razor and Sagan's balance – are vital in enabling us to think clearly about the nature of the paranormal. If we have a grasp of these three ideas, we have gone a long way to grasping the skeptical approach. However, by themselves they do not focus on particular paranormal claims. To do this, we need to examine particular claims. Later in this chapter, we shall look at a major – and sensational – claim about the paranormal, the remote viewing controversy. It involves professional researchers and a range of evidence which appeared to establish beyond any reasonable doubt that paranormal abilities exist, and indeed are accessible to everyone. As we look at the course the controversy took, we will see how skeptical tools can be applied to the paranormal claims, and can illuminate the outcomes.

It would be misleading to imply that there is only one skeptical approach. Like science, skepticism is a human activity, and so very difficult to categorise. We could look at Bertrand Russell's modest set of skeptical rules (1977a), or Carl Sagan's (1997) baloney detector, or the SEARCH algorithm proposed by Schick and Vaughn (2002). In addition, Wayne R Bartz (2002), a former psychology academic, has developed an approach he terms the CRITIC algorithm. This guides the researcher through different stages of evaluating a paranormal claim.

All of these approaches work. All of them enable us to sharpen our investigatory skills so that we can better evaluate paranormal claims. However, the three principles are especially valuable: they are the simplest of the approaches, and they focus upon the absolute heart of skeptical thinking.

WHAT THE SKEPTICS *REALLY* DO

So far, we have treated skepticism on its own terms, as the responsible scientific investigation of paranormal claims. On the other hand, the reader may have noticed that the skeptical principles, and the other approaches, do much more than this. What if, for example, an alternative explanation to a paranormal claim is that it is faked? This is not a scientific explanation, but it is probably more economical than a paranormal claim, thus invoking Occam's razor. It is time to ask: exactly what do skeptics actually do, since it is clear that they do not simply investigate scientifically.

In fact, skeptics do three things. First, they do indeed examine paranormal claims in scientific terms. However, they also explain claims of the paranormal in terms of human fallibility and error. As we shall see, they point to human misperceptions and mistakes in many paranormal areas. These are not scientific explanations, though they might be called 'natural' ones. Finally, the skeptics also explain claims of the paranormal in terms of human deception and self-deception. In the most extreme cases, the skeptics point to cases of outright fraud among paranormalists. Now, these latter types of explanation are not scientific, and so large parts of what the skeptics do actually falls outside their formal brief.

Does this mean that the skeptics are being deceptive or dishonest? Not really, both CSICOP (now CSI) and the Australian Skeptics have additional clauses which cover these extra activities. CSICOP refers to 'the use of reason in examining important issues', while the Australian Skeptics want to draw attention to 'natural and ordinary' explanations. These rather neatly cover explanations in terms of human error, deception and self-deception. However, they do mean that the role of science in the skeptics' explanations is not quite as large as their aims suggest. It is worth looking at examples of each of these types of explanation, as it helps to bring to life exactly what is going on.

When the skeptics' approach involves science, it falls into one of two categories. Often, skeptics use scientific knowledge to explain a paranormal phenomena. A rather elegant example of this approach is to be found in Barry Williams' article 'UFO is IPO' (Williams 1993). In this article, Williams describes how he was approached by a highly excited young man who claimed to have videotaped a UFO and was anxious to claim a substantial reward. Naturally, Williams investigated the claim. The UFO pictures turned out to be a vague, bright moving blob on a dark background. Williams worked out exactly where the camera had been pointing when the film of the 'UFO' was made. Then he consulted knowledgeable astronomers, and concluded that the object on the videotape was in fact Venus. An important point in the explanation was that at no time was anything like Venus shown on the videotape, yet it was prominent in the sky at the time that the film was taken. In this case, Williams was using astronomical knowledge of what was visible in one part of the sky to explain what had been videotaped.

The second type of scientific approach is when a scientific *method* is used in an analysis. For example, the Victorian skeptics, in Australia, have organised a series of trials of people who claim to be able to dowse for water. The essence of their method is that of a controlled experiment. The dowsers were kept in ignorance of which containers held water and which held sand until they had tried to dowse which were which. The skeptics were then able to show that none of the dowsers were able to achieve results better than chance (Sceats 2000; Australian Skeptics 2003).

Both the Williams approach, using scientific knowledge to investigate the validity of paranormal claims, and the Sceats approach, using an established scientific method, fall quite comfortably within the remit of the skeptics to scientifically investigate the paranormal. The other methods do not fall quite so easily within their goals, but are perfectly defensible. Let us look at how each of these work.

First, skeptics often explain paranormal claims in terms of human error and self-deception. It is very easy to think of ourselves as having infallible perceptions and reliable memories – as though we are really cameras linked to computer hard drives. In fact, as we shall see, neither our perceptions nor our memories are really all that reliable. One consequence of this is that skeptics are especially interested in the work of psychologists and magicians. In their very different ways, these two groups of people are interested in the ways in which we misperceive and misremember what we experience (e.g. Neimark 1996). They can then explain much of what is claimed to be paranormal in terms of the lack of reliability of our senses and memories. For example, massive sightings of UFOs across the US yielded clear cut statements of how some of the UFOs had windows, exhausts and were shaped like inverted saucers (Schick and Vaughn 2002, pp. 43–5). In fact, the 'UFO' was a disintegrating Russian spacecraft, and all the details came from people's faulty perceptions and memories.

Finally, skeptics occasionally accuse paranormal proponents of fraud or hoaxing. In the first chapter, we saw the case of a young woman who paid thousands of dollars to have a 'curse' lifted. The bogus clairvoyant was later arrested (Davis 2005). Psychic surgeons – who claim to be able to operate on the human body and remove diseased tissue without using knives or causing incisions – have also been accused of outright fraud (Plummer 2004b [1986]). In addition, it seems clear that large numbers of the crop circles which spawned a major paranormal movement were in fact caused by a couple of tricksters (Hempstead 1992; Schnabel 1993).

From all of this, it seems clear that skeptics cannot and do not simply confine themselves to scientific investigation. Most skeptics are not scientists, and it is clear that many paranormal phenomena need a range of skeptical approaches to tackle them. As we can see, these mostly fall under the headings of error, self-deception and fraud.

Looking at the aims of skeptical organisations, it is also worth noting that the Australian Skeptics have an overriding commitment to truth. The last clause in their aims involves 'the quest for truth wherever it leads'. Presumably this means that if the evidence for the genuineness of some paranormal phenomenon became overwhelming, then the skeptics would acknowledge this. It is a brave commitment, by which the skeptics align themselves with truth rather than simply with natural explanations.

In the next section, we will look at a well-known controversy which pitted paranormalists against skeptics. As we shall see, the final explanation was not scientific at all, although scientific principles were certainly involved on the way.

THE REMOTE VIEWING CONTROVERSY

The journal *Nature* is one of the world's best and most prestigious scientific publications. Over the years, a whole range of important scientific reports have appeared in its pages. Perhaps the most important of these was in 1953, when James D Watson and Francis Crick reported the double helix structure of the DNA molecule (Crick and Watson 1953).

In 1974 a paper appeared in *Nature* which, potentially, was far more important than the work of Crick and Watson. Two American researchers, Russell Targ and Harold Puthoff (1974) reported on a series of experiments which appeared to overturn the scientific view of the universe. They reported that remote viewing, their term for ESP, had been shown to be possible under tightly controlled laboratory conditions. In the same issue, the editors of *Nature* published an editorial explaining why they had published such an unusual paper (*Nature* 1974).

Scientific controversies can be slow moving, with months passing between published comments. Even by these standards, the remote viewing controversy went at a snail's pace. In 1978 two New Zealand psychologists wrote a paper (Marks and Kammann 1978), reporting that they had failed to replicate Targ and Puthoff's experiments, and explaining what they thought had gone wrong. Three years later Puthoff and Targ (1981) wrote back, with an account of how they had improved their results, and they were critiqued by Marks the same year (Marks 1981). Meanwhile, another parapsychologist, Charles Tart (1980) had weighed in on the side of the Americans (Tart et al 1980). Finally, Marks and Scott wrote another critique in 1986, arguing that Puthoff and Targ had still not met their criticisms.

However, the authors did not confine themselves merely to writing for *Nature*. In 1977 Targ and Puthoff wrote a book (Targ and Puthoff 1977) arguing for the importance of their work, while in 1980 Marks and Kammann (1980) also wrote a book which was skeptical about the psychic generally, but added a great deal of background evidence on the Targ–Puthoff experiments. More recently Schnabel (1997) and Ronson (2004) have added a whole new dimension to the controversy, detailing secret military uses of remote viewing, and what was revealed. After Richard Kammann's death, David Marks produced a second edition of their book (Marks 2000).

What was the controversy about? To understand it deeply, and to see how skeptical concepts like the burden of proof, Occam's razor and Sagan's balance are involved, we must first look at the original experiments and see how they worked. In essence they appeared to be a series of carefully conducted experiments offering strong evidence for the view that large numbers of people, if not everybody, had some level of psychical ability. Further, the results were so extraordinary that there was only one chance in many billions that they could have arisen by chance alone.

The essence of the remote viewing experiments was this. A remote viewer, notably former police commissioner Pat Price, entered a locked room in the Stanford Research Institute along with one of the experimenters. The room was designed to screen out any radio emissions which might enable communication to occur. The other experimenter, usually accompanied by research assistants, visited a location chosen at random from a pre-selected list. Of course, neither the viewer nor the experimenter with him knew this location. At a pre-agreed time, the experimenter at the location would begin to explore it, and the viewer would begin to describe any impressions he was receiving. The experimenter in the room with him would assist him with questions and encouragement. The viewer's comments were recorded and transcribed.

After the travelling experimenter had returned from the site, the experiment would be concluded. The viewer would then visit the site, and would often feel encouraged, pointing out aspects of the site which matched what they said in the sealed room.

After a series of experiments had been concluded, transcripts of the tapes and a list of the sites were sent to a judge, who had no knowledge of the order in which sites had been visited. The judge visited the sites, and then attempted to match the sites to the viewer's descriptions. The judge would examine each site, and rank the transcripts in order of their resemblance to the site. The results of the first nine experiments were amazing. Purely by chance, one would expect that in only about one case out of the nine would the correct site be ranked first. In fact, the judge ranked the correct site first in seven out of nine cases!

In all, the Stanford experimenters carried out a total of seven series of experiments, with 54 individual remote viewings. Because of the precise specifications of the experiments, they were able to calculate the probability of the results arising by chance. All but one of the series of experiments were strongly significant statistically. Using a conservative calculation, the odds against these results arising by pure chance was assessed as 5.6×10^{-4}, or less than one in ten thousand. A less conservative calculation – using several judges – put the odds against at 8×10^{-10}, or less than one in ten billion. As Marks and Kammann comment in their book, 'If all these statements are true, the Targ–Puthoff effect is the single most significant and convincing discovery in over one hundred years of psychic research.' (Marks and Kammann 1980, p. 35)

Perhaps the most striking feature of the Targ–Puthoff experiments is that, apart from Pat Price, the remote viewers were not people who particularly believed that they had psychic powers. Yet nearly all of them were able to achieve highly significant results. The implication is that not only do psychic powers exist, and can be demonstrated by replicable experiments, but that most, if not all, of the population possesses them. Targ and Puthoff concluded their astonishing account with the words:

> Our observation of the phenomena leads us to conclude that experiments in the field of so-called paranormal phenomena can be scientifically conducted, and it is our hope that other laboratories will initiate additional research to attempt to replicate these findings.
>
> (Targ & Puthoff 1974, p. 607)

This hope was fulfilled by two researchers in the Department of Psychology in the University of Otago, New Zealand. David Marks and Richard Kammann were, in their own words 'inspired by the SRI RV reports' (Marks and Kammann 1980, p. 35) and set out to replicate them. Between 1976 and 1978 the New Zealand psychologists ran a total of 35 experiments which, as closely as possible, replicated Targ and Puthoff's work. One minor difference was that the remote viewers were all inclined to think they might have psychic powers. That should make little difference if, as Targ and Puthoff believe, everybody actually does possess them. Targets were randomly selected from around the vicinity of the University of Otago. As a matter of course, the researchers introduced a series of precautions:

All transcripts were undated, marked with a random code letter and carefully checked for any cues that could aid the judge, such as reference to
previous targets. No such cues were available in any of transcripts. (Marks
& Kammann 1980, p. 37)

The researchers noticed that when the remote viewers visited the sites, they
often showed great confidence that they had successfully viewed them. One
viewer remarked for example that 'If the judges can't match my descriptions
correctly, there will be something wrong with them' (Marks & Kammann
1980, p. 35). Then a number of judges – students, academics and others
– examined the transcripts and the locations, and ranked the transcripts in
terms of their resemblance to each location.

However, when the judges' rankings were actually analysed, Marks and
Kammann had a shock. Here is the result:

None – repeat none – of the results was statistically significant. In not a
single case did a judge do better than chance at ranking the transcripts – a
total of twenty sets of judgements and not a single significant result.
(Marks & Kammann 1980, p. 37)

Marks and Kammann found that everyone tended to blame them. 'What
a mess – the subjects blamed the judges, the judges blamed the subjects,
and inevitably all psychic believers will somehow blame us.' (Marks 2000,
p. 38)

What was going on? How did matters change from the spectacular
results of Targ and Puthoff and the great confidence of the remote viewers
to the complete failure of any of the New Zealand results to be statistically
significant at the end? After a good deal of thought, Marks and Kammann
concluded that they were seeing an example of what they call *subjective
validation* (Marks 2000, p. 41). This happens when a person is required to
match two experiences, and they tend to focus on the positive resemblances
between them, and to ignore any negative discordant factors. Again and
again, Marks and Kammann noticed how the remote viewers, visiting the
sites, picked out features which confirmed that they had 'seen something'
and neglected whole ranges of factors which indicated that they had not.
As Marks and Kammann put it:

In the context of remote-viewing experiments, the subject obviously hopes
to do well and produce an accurate description of the target. When she goes
to see the target site after finishing her description, *she tends to notice the*

matching elements and ignore the nonmatching elements. Equally when the judge compares transcripts to the target and makes a relative judgement, he can easily make up his mind that a particular transcript is the correct one and fall into the same trap . . .

(Marks 2000, p. 41)

Of course, subjective validation is an important part of many human experiences. It is one of the reasons that we have to take care to check our own beliefs all the time and always bear in mind that we may be wrong. However, it is not enough to explain the totality of what happened to remote viewing. It does not explain the powerful statistical significance of Targ and Puthoff's result compared to the complete failure of Marks and Kammann's work to replicate them. Something else is clearly at work here.

Marks and Kammann reviewed their experiments, wondering what they had done wrong. Another set of experiments carried out in the US, which had failed (Marks 2000, p. 45), turned their thinking in another direction. What if it was not their experiment that was wrong, but that there was something wrong with the work of Targ and Puthoff? With this in mind, they had a careful look at the original remote viewing experiments again.

Marks and Kammann focused on the cues which they had edited out of their transcripts. It seemed essential to edit out such cues. Comments such as 'the church we went to yesterday' could, if there were a number of them, bias the statistical results. Indeed, if such cues were scattered about in all the transcripts, and if the judges' list of targets was in the order they were visited, it might be possible for a judge to figure out what transcripts corresponded to what locations. In Targ and Puthoff's book, Marks and Kammann found two transcripts, and in both there were cues that had not been edited out. They set out to investigate this, but Targ and Puthoff refused to allow them access to the transcripts. A judge, Arthur Hastings, was more helpful in this regard, and they eventually found that the targets were listed in the order in which they had been visited, and the transcripts were unedited. Using this information, it is perfectly possible for a judge to match the transcripts to the targets perfectly, with no paranormal abilities at all. Marks, using the five transcripts he had obtained, was able to match them perfectly to the locations, even though he had not visited the sites: he did this purely on the basis of the cues.

This is an astonishing letdown from the promise of the original experiments. Marks and Kammann make the point that the quality of the transcripts was extremely poor. Wading through a mass of poor resemblances, the judges would have almost certainly used any cues they could find to match the transcripts to the sites, and this is how the remarkable results of the Targ–Puthoff experiments, apparently, were obtained.

The controversy trailed on for several more years in the pages of *Nature*. Targ and Puthoff conceded that their original experiments were flawed, but argued that later ones were better controlled and, in any case the influence of the cues could not have been sufficient to cause the results as they occurred. Another paranormal researcher (Tart et al 1980), also wrote to say that he had edited the transcripts and had them independently judged, and that the judges were still able to achieve the remarkable 'hit rates' of the original experiment.

Marks was not convinced. He and Scott (1986) pointed out that since some of the transcripts had been published with cues, it was invalid to use them again in experiments, and that since Targ and Puthoff had still refused to release all the transcripts, it was impossible to check the results properly.

Obviously, with the unpublished transcripts, it might still be possible to check the claims of both the parapsychologists and the skeptics here. The transcripts could be edited, and then given to judges. One set of judges might look at the unedited transcripts, another set at the edited ones. In this way, it would be possible to see if in fact the information in the transcripts alone could provide the results, or if it is the cues which do so. However, since the transcripts are mostly not released, that is impossible.

SKEPTICAL PERSPECTIVES

What should we make of all this? In the tradition of many disputes, the remote viewing controversy has dissolved into a mass of ambiguities and irrelevancies. Clearly, Marks and Kammann had to develop their own concepts of what was happening. Their explanations were that information leakage was occurring to the judges, and so the entire process was being thrown into doubt. The specific concept that seemed to explain the evidence was the 'cue'. These enabled the judges to assign remote viewing transcripts to particular places without any need for parapsychological processes. Using the more general skeptical concepts, it is possible to spell out a few clear points which point to a conclusion.

First, the burden of proof clearly lies with Targ and Puthoff. Although they do not say so explicitly, they do indicate that they accept the burden, in at least two ways. First, they produce spectacular statistics to support their claims and, second, they express the hope that other people will replicate their experiments and so verify them (Targ and Puthoff 1974). Clearly, these researchers were aware of the astonishing nature of the claims they were making, and of the obligation of researchers to produce evidence which would justify them.

The explanation used by Marks and Kammann clearly involves the use of Occam's razor. Marks and Kammann argued that the 'cues' – clues to the order in which sites had been visited – provided sufficient information for the results, without any recourse to extrasensory perception. Indeed, Marks himself was able to achieve 100 per cent accuracy in allocating some transcripts to sites without visiting any of the sites himself, purely on the basis of the cues. From Occam's razor, it follows that if a straightforward natural explanation exists, there is no need for the spectacular paranormal explanation: Targ and Puthoff's claims are not justified.

However, Targ and Puthoff, and also Charles Tart, made something of a comeback, arguing that careful editing of the transcripts, to remove all the cues, still enabled judges to rank the transcripts in an order which was significantly better than chance. Marks wrote back, criticising these claims and pointing out that the transcripts were still not accessible to the researchers generally in the area.

Using Sagan's balance, we have to ask whether the evidence, after all the arguments, is sufficient to convince us that remote viewing is other than a natural process of information leakage. Hume, of course, would say no, and that no evidence could possibly be sufficient. However, we can examine the evidence and see whether it is remarkable. Apparently, in the original experiments, the cues were not edited out of Targ and Puthoff's transcripts, and in addition, Marks seems to have established that documents with cues in them can easily be linked to the sites visited. This certainly renders the evidence very unremarkable indeed. What of the claim that after editing, Targ and Puthoff's transcripts could be ranked by judges? This clearly puts us back at stage one. The original experiment has not established the level of evidence required, and even then Targ and Puthoff were hoping that other people would replicate their work with similar results. Hence, unless other people can produce similar levels of significance for remote viewing experiments – with the cues carefully edited out – then we are entitled to say that the evidence does not meet the requirements of Sagan's balance, and so does not convince us that anything is at work here other than human error.

In this example, we have looked at a claimed case of the paranormal in some detail. We have seen how the use of the three concepts helps us to see what was happening, and enables us to come to an assessment of the evidence for remote viewing. In this case, it looks is if the evidence is not enough to tip Sagan's balance. Could it have been? In my view, the answer is yes. If Marks and Kammann had replicated Targ and Puthoff's experiment and obtained similar results, then there would have been a very good case for taking the evidence seriously. Then, if another set of independent researchers had made similar findings, the phenomenon of remote viewing might have been well established.

It is worth noting that the sequence of events here often appears in parapsychology. Researchers begin with high hopes and claim that an important breakthrough has been achieved, or is about to be achieved. Then, slowly, over a much longer time the results are shown to be inconclusive and interest lags, until the next sensational development occurs. A sad example of this appears in Deborah Blum's *Ghost Hunters* (Blum 2007). This book tells the story of how a brilliant group of nineteenth century researchers, including William James and Oliver Lodge, set out to investigate mediumship scientifically. They began with great hopes, but as the decades went by, they found themselves essentially where they started. The latter part of Blum's book resembles a tragedy, as the brilliant investigators age and begin to die without achieving their objective. This pattern has been repeated many times in parapsychology. What the field needs to be accepted is a single repeatable experiment, one clear means of demonstrating that paranormal powers exist. Until that happens, the field will remain outside science.

WHAT EXACTLY IS SKEPTICISM?

We have now looked at some of the key principles of skepticism. We did this both with the chapters on science and on the paranormal. With both of these entities, we then went on to look at their place in the modern world. With science we saw that it is a multi-billion dollar colossus, based in universities, government labs and business organisations, which has had massive effects on all aspects of our lives. With the paranormal, we saw that it is outside the normal structures of research, but is supported by a varied set of social mechanisms such as the cult or the client–practitioner network. Further, there are processes in the normal human mind which predispose us towards paranormal beliefs.

What about skepticism? Can we distinguish some features of skepticism and the way that it is supported in modern societies? There are two tempting ways to look at skepticism, and both appear to be wrong. First, one can see skepticism as some kind of academic movement. After all, some prominent skeptics – such as Paul Kurtz, Susan Haack and Richard Wiseman – are academics. However, a slightly closer look shows the falsity of this. First, there are no academic organisations devoted to skepticism. There is no Skeptical Research Centre on any university campus, nor is there a refereed *Journal of Skeptical Studies*. Further, many prominent skeptics are not academics. Examples are Martin Gardner, James Randi, Barry Williams and, indeed, David Hume himself: he wanted to be an academic, but was not able to be appointed. Therefore, it is simply incorrect to regard skepticism as academic.

Another tempting possibility is to regard skepticism as a religiously based movement. Instead of being associated with a particular religious movement, as creationism is with fundamentalist Christianity, it is alleged that skepticism is associated with atheism. The prominent parapsychologist George P Hansen (1992) has made this allegation at some length in a critique of CSICOP.

Without doubt there is some truth in the allegation. Many prominent skeptics are indeed humanists or rationalists, and David Hume, the intellectual father of modern skepticism, was a strong critic of religion. At the same time, many skeptics are not atheists or humanists, starting with Martin Gardner. Further, there is nothing necessarily atheistic about seeking to test claims of the paranormal. After all, the nature of the paranormal means that it is a this-worldly phenomenon, and there is nothing atheistic about examining this-worldly claims with the highly effective this-worldly tool of science. Unless their religious beliefs are directly involved, there is nothing to prevent any religious person taking part in this endeavour, just as enthusiastically as any atheist. Therefore, it does not seem true that atheism is the prime motivator of skepticism.

If skepticism is not academic and not atheistic, what exactly is it? Perhaps the best answer is a category we have already seen used by sociologist Erich Goode: it is a grassroots movement. Other sociologists use the term 'social movement', and we will adopt this. What is a social movement? It is a grouping together of people in support of a common goal or concern, which sociologists term a 'frame' (Buechler 2000). Of course, with the advent of the internet, the organisation of social movements has become much more widespread.

What is the frame which mobilises the skeptics? In general, as we saw earlier in this chapter, it is the concern that paranormal beliefs – beliefs plainly contradictory to scientific knowledge – are widespread in society, and becoming more and more influential. This perception brings together a wide range of people who might otherwise have little in common. Doctors, for example, might become concerned about the use of unproven alternative health remedies. Atheists might deplore the woolly-minded non-materialistic nature of many paranormal beliefs. Scientists might be shocked to find their findings being given less weight than those of a pseudoscientist. Academics might be concerned by the lack of scientific literacy in the general population, and so on.

My own recruitment into skepticism occurred in the 1980s, when the Queensland state government was dominated by the eccentric fundamentalist Johannes (Joh) Bjelke-Petersen. A fundamentalist Minister of Education began a vigorous push to have creation science included in the school science syllabus. A panel of science teachers was convened – liberally seasoned with fundamentalists – to decide how (not whether) creation science should be taught in Queensland schools (e.g. Knight 1986). Many people who might normally speak out against these developments were silenced. I know one lecturer at a College of Advanced Education who spoke out strongly against creation science, and who was told by the Principal of the college to stop his criticisms. Apparently orders had come down from higher up.

In this dire environment, the Australian Skeptics were one of the few groups to became heavily involved. They risked a large part of their annual budget to finance a book, co-edited by myself and Ken Smith (Bridgstock and Smith 1986), which criticised all aspects of creationism. The book also revealed some rather ludicrous financial losses by the main creationist organisation in Queensland (Bridgstock 1986b). The Minister of Education backed away from his support of creation science, and the crisis gradually subsided. My own reaction to the skeptics' activities is not mere general support, it is also one of profound gratitude.

Skepticism, therefore, seems to be a grassroots social movement, not based in academia or in science, though it has some supporters in both areas. Although skeptics do a good deal of work exposing fakes and the self-deluded, it is also clear that they have not thought out their own position very carefully. This book is, in part, aimed to correct that problem by clarifying what skepticism is about, and why it is worth supporting.

Since the first draft of this book was written, CSICOP has acted in ways which suggest that they are seeking to clarify their goals and activities. They have shortened their name to the Committee for Skeptical Investigation, and broadened their goals to include the 'investigation of controversial or extraordinary claims'. (Frazier et al 2007, p. 6) Kurtz (2007) points out that this does not mean that paranormal investigation has been abandoned, only that a broader range of topics can be looked at. We will consider this change later in the book.

LAUNCHING PAD

For those interested in being a skeptic, it is necessary to understand some of the basic principles involved. This book has introduced three basic principles, and Caso (2002) outlines exactly why they are justified. There are other approaches to skepticism, and it is useful to look at Wayne Bartz's CRITIC approach (Bartz 2002) or Schick and Vaughn's SEARCH algorithm (Schick & Vaughn 2002). There are other approaches as well. Bertrand Russell, the great philosopher, espoused a series of simple skeptical rules, which are gently stated but, in their own way, quite devastating (Russell 1977b). There is also the 'baloney detector' of Carl Sagan (1997a). It is a good idea to become acquainted with one of these approaches, and preferably more.

Blum's book is a sad evocation of the optimism with which Victorian researchers began to study the paranormal, and the growing realisation, as the decades went by, that they were getting nowhere (Blum 2007).

How do people become skeptics? My own involvement has already been mentioned. Michael Shermer, describes in the first chapter how he became a skeptic during a long-distance cycle race (Shermer 1997), while Karla McLaren, a former New Age guru, touchingly describes her conversion to skepticism, and the renouncing of her previous beliefs (McLaren 2004). Susan Kawa (2003) provides a contrasting account. For a good description of how the skeptical movement came into existence, it is worth looking at the series of essays in Kurtz (2001a). They clearly show why the movement came into existence, and what the early founders thought and felt.

In the second half of the book, having delineated the broad landscape, we will descend to the battle plain. We will look at the specific tools and weapons that the skeptics use to analyse paranormal claims, and how the paranormalists respond to them. In looking at these skirmishes, we will gain a good idea of what is happening, and how the battle is going.

5

Bringing skepticism down to earth

IN THE PREVIOUS chapter, we looked at the key principles of modern skepticism. Between them, they provide a foundation for a skeptical outlook on the paranormal and, as we shall see, on the world generally. However, if the reader tries to apply these concepts directly to the paranormal, there is likely to be disappointment. The ideas are far too general to apply to most paranormal claims. For example, Sagan's balance states that the evidence must be sufficiently strong to justify an extraordinary belief.

But how does one decide exactly what the evidence is, or whether it is strong enough? The general statements give little guidance. Clearly, another step is needed between the general skeptical principles and the actual paranormal claims. For any claim, the skeptic has to understand the main ideas involved in the claim, the main skeptical ideas that can be brought to bear, and the evidence which is involved. There is no shortcut or easy way, although some approaches are better than others. We will call this new set of ideas 'intermediate concepts' as they stand between general skeptical principles and the amazing range of paranormal claims.

Paranormalists who approach me are often disappointed. They proffer some sort of evidence ('My friend saw a UFO!', or similar) and expect me to be able to demolish the claim there and then, or else to concede that the claim is justified. In fact, of course, this is impossible. The whole point of skepticism is that it is essential to look carefully at the evidence, and to understand what it means. Skepticism is not a quick fix. It is much more like a road map, showing how to examine claims. However, just looking at the road map gets you nowhere: you have to go out and make the journey. This chapter is aimed at enabling you to make the transition between the map – the skeptical principles – and what is really happening in the world. These intermediate concepts, therefore, are what make skepticism work.

THE INTERMEDIATE CONCEPTS

What are these intermediate concepts that stand between skeptical principles and the actual analysis of paranormal claims? For example, we are considering a particular paranormal claim; perhaps someone has seen a UFO. By itself, the general principles of skepticism are not much help. We can say that the burden of proof lies on the claimant. However, we do not have any clear guidelines as to exactly how that proof might be accomplished. Occam's razor tells us that we should minimise the number of explanatory entities. Fair enough, but if all we have is someone's claim that they saw a UFO, how do we work out what other explanatory entities there may be? Sagan's balance tells us that extraordinary claims require extraordinary evidence, but how do we go about working out exactly how extraordinary the evidence really is? In each case, we find ourselves without a way of applying the principles to specific cases. Here is where the intermediate concepts come in.

Broadly, there are two types of intermediate concepts. One type involves the assessment of the strength of evidence. For example, in the case of the UFO claimant, there is a good deal of evidence about the reliability of eyewitnesses. People often see things that are not there, or incorrectly report things that they did not see. In many cases, this sort of knowledge can strongly affect our assessment of paranormal claims. If it seems likely that someone is reporting incorrectly, or is seeing something that is not there, then the evidence becomes much less extraordinary, and Sagan's balance tips against the acceptance of the claim. On the other hand, if we apply these sorts of concepts and the sighting turns out to withstand the examination, then we have to conclude that the evidence is a great deal stronger. Skepticism can work both ways: if evidence is weak, paranormal claims can be discarded, but if the evidence is genuinely strong, then the claims might end up being verified. And our view of the universe will never be the same again.

The second type of intermediate concept involves natural explanations for what is claimed to be paranormal. We have already seen a simple example of this. It came from Barry Williams (1993). He was able to show that a 'filmed UFO' was in fact the planet Venus, and so explained the sighting.

In this case, the intermediate concepts were the position and luminosity of the planet Venus. It enabled Barry Williams to formulate an alternative explanation. By Occam's razor, since a natural explanation exists, one has no need of a complex idea involving aliens and their flying machines. So, in this case, Barry Williams used an explanatory intermediate concept to explain the film, and Occam's razor eliminated the claim.

Clearly the intermediate concepts are likely to vary from one paranormal area to the next. Potentially, there are thousands of them. Skeptics have compiled directories of them, such as Kelly's *The Skeptics Guide to the Paranormal* (2004) and the online *Skeptic's Dictionary* (Carroll 2008). Sometimes, the investigator has to think up new ones. There are a few ideas that are of great importance, and we will look at them here. We will start with a series of ideas which enable skeptics to examine carefully eyewitness claims.

THE RELIABILITY OF EYEWITNESSES – VISION AND MEMORY

As we have already seen, David Hume's famous attack on the credibility of miracles focused on eyewitnesses. His conclusion was that no number of eyewitnesses can be sufficiently convincing for a 'wise man' to conclude that a miracle has taken place (Hume 1980 [1748]). Still, in many fields of the paranormal, there is heavy reliance on the testimony of eyewitnesses. Large parts of ufology, ghost hunting, poltergeist reports and sightings of the Loch Ness Monster depend on the accounts of eyewitnesses. We are often assured that the people making the claims are reliable or reputable. But how much weight can be given to their evidence?

In fact, a whole range of evidence strongly suggests that even honest people, strongly convinced of the truth of their accounts, are not very reliable. When we think about whether to trust eyewitnesses, there are two issues, which are very closely connected. One is the nature of human vision and other senses. The other is the nature of human memory. It is terribly easy to make assumptions about the nature of human eyewitnesses, and the nature of human memory. The human eye is, in principle, rather like a camera, and so it is easy to assume that eyewitnesses record events they saw as reliably as a movie camera. Equally, because we are all familiar with the idea of computer memory, it is easy to assume that what we memorise is in some way stored in a memory, like a hard disk, and can be recalled at will.

Elizabeth Loftus, an American psychologist, is foremost in demonstrating the error of such ideas. In studying eyewitnesses, Loftus and a colleague carried out two key experiments. In the first she showed a film of a car crash, and then asked questions about what the viewers had seen. She found that the way that the questions were phrased had a great effect on what people reported seeing. For example, people were asked the speed at which cars appeared to be going if they 'hit' each other and if they 'smashed' into each other (Loftus and Palmer 1974). Those who were asked about the cars smashing into each other estimated, on average, that the cars were travelling 7 mph (11 km/h) faster than those who were asked about the cars hitting each other. The film did not show any broken glass. However, when people were asked whether they had seen broken glass a week later, the ones who were asked about the cars smashing into each other were markedly more likely to say that they had seen broken glass (32 per cent compared to 14 per cent).

In an equally important experiment (Loftus, Miller and Burns 1978), Loftus and her colleagues showed people a series of slides of a motor accident. Some people saw an accident involving a 'stop' sign, while others saw it involving a 'yield' sign. Then they were asked questions which either agreed with the slide, referring to the same sign, or which disagreed, referring to the other type of sign. Then they were asked to select the original slide they had seen. It turned out that while 75 per cent of the people who had received consistent information picked the right slide, only 41 per cent of those who had received inconsistent information picked the right one.

Loftus argues that this shows that memory does not work like a simple storage system, but that it is a dynamic process, intimately connected to what we are looking for (Loftus 1979). What is more, when people search their memories to give an account of what they have seen, they use any fragments or clues that are available to them, including those from phrases, subsequent events and what they know about these types of events. This makes it extremely dangerous to rely upon the unsupported memories of eyewitnesses.

The implications of this are disconcerting. It suggests, as Loftus thinks, that our memories are constantly being reconstructed, and that we use information quite indiscriminately to hold our memories together. Thus, we can claim to have seen things that did not occur, or not seen things that were actually there. And, in consequence, our memories can be easily polluted by suggestions from other sources.

Psychologists have shown exactly how poor we are at seeing and remembering things, even if we see them every day. As French and Richards (1993) summarise, Americans are unable to recall information about their nickels, British people cannot recall details of their ten-penny coins, and also cannot reproduce the letter and number configurations on their telephones.

All this suggests that we should take great care when we accept the testimony of eyewitnesses on almost any subject. Ufologists are often fond of using eyewitness testimony with profuse comments on the reliability and honesty of the people themselves (e.g. Good 1987). However, what the experiments of Loftus and her colleagues show is that no matter how honest and honourable people are, being human makes it extremely hard for us to testify reliably on what we have seen.

There is a concrete example of this, which skeptics Schick and Vaughn have collated into a single embarrassing story. We will summarise its main features, as it shows clearly the problems of relying on eyewitnesses for accounts of amazing claims. As Schick and Vaughn describe:

> Records from the North American Air Defence Command (NORAD) and other evidence show that at the time of the UFO sighting, the rocket used to launch the Soviet Zond 4 spacecraft reentered the atmosphere, breaking into luminous fragments as it sped across the sky. It zoomed in the same southwest to northeast trajectory noted by the witnesses, crossing several states.
>
> (Schick & Vaughn 2002, p. 44)

The story becomes surprising when the details of the witnesses' reports are considered. Clearly, if the details of the reports agree with each other, and are in accord with the re-entry of the rocket, then we might regard the case as a good piece of evidence in support of the reliability of the eyewitnesses. On the other hand, if the reports are inconsistent with what actually happened (as Loftus found in her experiments), then we might conclude that eyewitness reports can be terribly misleading. So what did the eyewitnesses report?

> In Tennessee, three educated, intelligent people (including the mayor of a large city) . . . observed orange-colored flames shooting out behind it, with many square-shaped windows lit from inside the object.
>
> (Schick & Vaughn 2002, pp. 43–4)

This is not all. In Indiana six people saw the object. And what did they see?

> Their report to the Air Force said that it was cigar-shaped, moving at tree-top level, shooting rocket-like exhaust from its tail, and it had many brightly-lit windows.
>
> (Schick & Vaughn 2002, p. 44)

But in Ohio, they saw things rather differently.

> But they said that they saw three luminous objects, not one. One of these witnesses used her binoculars to get a good look at the UFO. She submitted a detailed report to the air force that said the objects were shaped like "inverted saucers", flying low and in formation, silently cruising by.
>
> (Schick & Vaughn 2002, p. 44)

The differences between the accounts are quite astonishing. And the clear statement of several apparently reliable witnesses does seem to show that people do see things that are simply not there. The Zond space launcher had no windows, and so they were clearly not illuminated from the inside. What is more, the witnesses' accounts also vary regarding the shape and number of objects.

The conclusion suggested by Loftus's experiments, the cases recounted by other psychologists and the startling disparities in the case of the Zond launcher all point to the conclusion that the memories, even of honest and well-intentioned people, cannot be regarded as powerful evidence, unless they are corroborated. Thus, eyewitness evidence on its own cannot reach the levels of extraordinary strength which Sagan's balance would require. It might be possible to have a series of eyewitnesses, who are kept quite separate from each other after the event, and who would give reports of the event that agree closely with each other's reports. If there were sufficient eyewitnesses, of sufficient probity and in close agreement, then we might conclude that this is strong evidence. The normal run of eyewitness testimony, however, would not be sufficient to be convincing.

Do these doubts about memory carry directly into memories of the paranormal? They certainly do, and devastating evidence has been found. As far back as 1887 psychic investigators Hodgson and Davey held fake seances, and asked the unsuspecting attendees to write about what they had seen. According to the researchers, accounts erred about major events, such as a key person leaving the room and how many people were present (Hodgson & Davey 1887). More recently, Richard Wiseman and his colleagues have also organised fake seances with similar results (Wiseman, Smith & Wiseman 1995): stationary objects were reported by many people as having moved. Modern researchers, however, have gone further. Psychological tests show who believes, and does not believe in the paranormal, and these prior beliefs have a marked influence on the accuracy of what is seen the seance room (Wiseman & Morris 1995; Wiseman, Smith & Wiseman 1995). The conclusion seems justified that

... there is now considerable evidence to suggest that individuals' beliefs and expectations can on occasion lead them to be unreliable witnesses of supposedly paranormal phenomena. It is vital that investigators of the paranormal take this factor into account when faced with individuals claiming to have seen extraordinary events.

(Wiseman, Smith & Wiseman 1995, p. 32)

We have already seen, in Chapter two, how psychologist Susan Clancy found that people gradually came to the conclusion that they had been abducted by aliens, and indeed developed memories of this event. In recent experiments, Whitson and Galinsky (2008) have shown that people experiencing a minor loss of control are significantly more likely to develop superstitions, perceive conspiracies and see patterns where there are none. In addition, Wilson and French have found that people with paranormal beliefs are more likely to have false memories of news items (Wilson & French 2006). There seems to be a rich lode of research for psychologists in this, and an endless source of concern for skeptics.

Obviously, the longer the time lapse between an event and when it is recounted, the less likely it is to be accurate. A recent example of this was the memory of President George W. Bush over the 9/11 terrorist attacks. The President gave at least three accounts of how he had heard about the attacks, and what he did. Conspiracy theorists pounced upon this, claiming that it showed that the President knew about the attacks in advance, or that he was in some way part of a conspiracy (Greenberg 2005). However, the truth is much simpler and much more human. President Bush has a normal human memory, and this means that events are constantly being recreated in his mind, so that inaccuracies creep in. And the same applies to everyone. We have to be very careful when using eyewitness accounts, and even more careful when there is a danger that they have been reprocessed in human memories. We should check different accounts, if they exist, and also try to find other evidence. This may show that the eyewitnesses were perhaps wrong, or it might strengthen their claim to be believed.

REPLICATION – A POWERFUL TOOL FOR SCIENTISTS AND SKEPTICS

We have already seen, in our look at science, that replication of experiment and observation is an important feature of many scientific fields. We also saw why. Theories are based on available evidence, and if the evidence is wrong, the theories are doomed to be wrong too. Therefore, it follows that wherever possible, scientists will want to repeat important experiments and observations reported by other scientists. After all, any experiment can be wrong, any observation can be an error. Therefore, it is quite understandable that key experiments and observations will be replicated.

In essence, in a replication, another scientist carries out, as nearly as possible, the same piece of work as was done before. The aim is to see whether the result is reliable. That is, will the same results be observed? Sometimes replicated results are the same as the original, sometimes they are quite different.

Replications are not routine in science. They normally occur only when the reported results are of great importance, and are usually controversial as well. Thus, when Michelson and Morley in 1905 reported that the speed of light seemed to be the same regardless of the direction of travel of the Earth, it created a sensation and led to many replications (Gardner 1957, pp. 84–5). The findings were upheld. On the other hand, when Pons and Fleischmann claimed to have initiated cold fusion in the laboratory, attempts at replication mostly failed, and the claims were eventually disbelieved (Close 1992).

Replication sounds like a clear and simple way to find out if results are correct. However, it is less certain than it sounds. Consider a case we have already looked at: remote viewing. In the previous chapter we saw how two New Zealand psychologists replicated the remote viewing experiments of Targ and Puthoff. To their amazement, they found that they were unable to reproduce the results, even though both their viewers and the judges felt confident that remote viewing had taken place.

After a great deal of thought, Marks and Kammann concluded that their experiments were not working because Targ and Puthoff had left 'cues' – comments about other remote viewing sessions – in the text of the transcripts, and these could explain the remarkable results. If Marks and Kammann had left the cues in their transcripts, it is perfectly possible that they would have achieved results similar to those of Targ and Puthoff. That is, they would have found that remote viewing exists.

Clearly, in many areas of the paranormal, replication is simply not feasible. How does one replicate a UFO sighting, for example, or a dramatic dream in which one 'psychically' foresees the future? On the other hand, it is always worth bearing the idea of replication in mind, because sometimes it can occur unexpectedly. Decades ago, I remember a rather sensational television story, in which an aircraft passenger had filmed a UFO from an aircraft porthole. The movie – shot through the porthole – showed a blurred disk apparently coming up to the plane, then zooming away with tremendous speed.

This was a sensational piece of news. However, a couple of reporters from the program investigated further. They replicated the passenger's movie, with the plane on the ground, and found that they, too, could film the UFO. The aircraft's portholes were very thick, convex glass and the 'UFO' was in fact the end of the plane's wing, greatly distorted. Small movements in the position of the camera apparently showed the 'UFO' advancing and receding at great speed. Amazingly, a 'UFO sighting' had been replicated!

We should also note that replication fits in very neatly with our characterisation of science in Chapter one. We saw there that the current workings of science are based on a community of investigators who consider the phenomena and decide what is established and what is not. Thus, the solitary genius labouring away in a laboratory is not how science actually works. Sometimes scientists might seem to be solitary, but they are part of a network of communication and decision, and it is the judgement of the scientific community which decides what has been established and what has not.

This is why replication is so very important in science. If the knowledge is communal, then knowledge held by one person is not fully accepted. It requires accessibility by all members of the community. This means that an astonishing experimental result will probably not be accepted by the scientific community until it has been replicated, preferably by scientists who are skeptical about the results. In this way, replication is one of the keys by which experimental results are accepted into the scientific community's storehouse of knowledge. It shows that the results are reliable.

Why should we bear replication in mind whenever considering any paranormal claim? The answer comes from Sagan's balance. Paranormal claims need strong supporting evidence, and normally a single observation or experiment would not be sufficient for a reasonable person to be convinced. Therefore if paranormal claims are to be accepted, we really want replications, or something equivalent, to be performed. If they are not, the chances of acceptance are very slight.

THE AMAZING COINCIDENCE

Our next topic is one of the main ways in which people are recruited to believe in the paranormal. This is the occurrence of an amazing coincidence, where something happens which seems so astonishing that there is no other explanation except the paranormal. Let us look at an example. Imagine that you travel to work by bus. While in the bus, looking out of the window, you remember Craig, a friend of yours from many years back. You wonder where Craig is now, and what he is doing. You have not thought about Craig for several years. Then your bus reaches its destination. You climb off the bus – and there is Craig, your friend!

Of course, you talk to Craig. You want to know what he is doing there. You find that his reasons are totally mundane: perhaps his car is out of action, and he is taking a bus to work. Still, the impact of the experience is likely to be considerable. You will probably feel astonished. If you do some calculations, you are still likely to be amazed. Perhaps you have not thought about Craig for ten years. That is a total of 3650 days (ignoring leap years). On the very day you happen to think about him, there he is. This means that there is only about a one in 3650 chance that this could have occurred purely by coincidence. It seems perfectly reasonable for you to conclude that something paranormal is at work. Many people have these types of experience. You might have been thinking about a person when the phone rings – and there they are, on the phone. Or, you are wondering about a person you have not seen for many years, and a letter from them arrives.

Before we begin to consider the matter skeptically, we should note that even more amazing events have occurred. Compared to these events, your experience with Craig pales into mundanity.

The psychologist Carl Gustav Jung was amazed by some of the accounts of coincidences he encountered. Here is an example of one that he noted:

> A young woman I was treating had, at a critical moment, a dream in which she was given a golden scarab. While she was telling me this dream I sat with my back to the closed window. Suddenly I heard a noise behind me, like a gentle tapping. I turned around and saw a flying insect knocking against the window-pane from the outside. I opened the window and caught the creature in the air as it flew in. It was the nearest analogy to a gold scarab that one finds in our latitudes... which contrary to its usual habits had evidently felt an urge to get into a dark room that this particular moment.
>
> (Jung 1960, p. 438)

In this case, Jung took advantage of the amazing event. The young woman had been proving very difficult to treat, and Jung recounts how he held out the insect to her, saying 'Here is your scarab'. Apparently the shock was so great that treatment went much better thereafter.

Jung developed a whole theory based on these coincidences, which in his view were 'connected so meaningfully that their chance occurrence would represent a degree of improbability that would have to be expressed by an astronomical figure'. (Jung 1960, p. 21)

Perhaps the most astonishing event was published in the London *Times* of May 1974. A competition was held for the most amazing story involving coincidences. There were two thousand entries, and the winner was a Mr Page. He told how he had been a soldier in 1940, and his long-awaited wedding photographs had been opened by another soldier, and this was not really surprising because their names and numbers were very similar. He explains:

> His being Pape No. 1509322 and my name being Page No. 1509321. This mix up in the mail being frequent until I was posted to another Battery. Some time after the war had ended, I was employed as a driver with London Transport at the Merton depot, Colliers Wood, S. W. London. One particular pay day I'd noticed that the tax deduction was very heavy, and duly presented myself to the Superintendent's office... Imagine my amazement when I discovered that my wages had been mixed up with a driver who had been transferred to the garage, not surprising when I found out that his name was Pape, yes the very same chap... the weirdest thing of all, our PSV licence numbers were – mine 299222, Mr Pape 299223.
>
> (Tutt 1997, p. 211–2)

Several eminent thinkers believe that there is more than coincidence in accounts like these. Jung developed a theory on the matter, as did writer Arthur Koestler (1974) and one of the founders of oceanography, Alister Hardy (Hardy, Harvie and Koestler 1973). Clearly, if these eminent thinkers are to justify a theory that 'something is going on' behind the coincidences, they have the responsibility of establishing it. The burden of proof is upon them. Can they justify their belief?

At least one eminent thinker who pre-dates any of those impressed by coincidences has a different view. Francis Bacon was a politician turned philosopher who was brutally dismissive of coincidences. In his view, talk of them should be despised, because they 'have done much mischief.' In his view, people give talk of such coincidence credit for three reasons:

> First, that men mark when they hit, and never mark when they miss; as they do generally also of dreams. The second is, that probable conjectures, or obscure traditions, many times turn themselves into prophecies; while the nature of man, which coveteth divination, thinks it no peril to foretell

that which indeed they do but collect... The third and last (which is the great one) is, that almost all of them, being infinite in number, have been impostures, and by idle and crafty brains merely contrived and feigned, after the event past.

(Bacon 1625, p. 44)

Bacon's first reason is that people notice amazing coincidences when they happen, but not when they might happen and do not. Then he seems to be saying two things in one sentence. The first is that vague guesses and traditions can be regarded as amazing prophecies because, like Hume, Bacon believes that people crave divination, and also that conjectures or traditions might turn themselves into prophecies. Perhaps someone might notice that an event resembles an older prophecy, or might act to make such an old prophecy come true. Finally, Bacon thinks there is a great deal of faking and fraud.

The idea that a prophecy may be self-fulfilling has an important application in medicine. As we shall see, if someone believing in a treatment is given the treatment, they may often become better, even if the treatment is wholly ineffective. This is the placebo effect, which we will encounter later.

Bacon's view was that, in looking at the reasons for astonishing coincidences, we are likely to find that fraud was the most important. This may not be the case. Fraud may be important, but another of Bacon's reasons is vital. This is the fact that people notice hits but not misses. To understand why this is important, we should look at the work of two New Zealand psychologists, David Marks and Richard Kammann (Marks 2000). These researchers, whom we have met in analysing the case of 'remote viewing', have put together an alternative – non-paranormal – theory of amazing coincidences.

Bacon says that people tend to notice 'hits', that is, when coincidences occur, and not 'misses' when they do not occur. But how are we to know when a coincidence could occur but did not? This is the contribution of these researchers. Imagine, say Marks and Kammann, that we can recall at the end of an ordinary day, a total of one hundred distinct events, if we are asked about them. These might be experiences like a dream, a news item, something that happened at breakfast and so on. Then they go on:

The first event can be paired with each of the ninety-nine others. The second event can be paired with each of ninety-eight others (because it has already been paired with the first event). Proceeding in this way we can see that the total pairs is given by $99+98+97+96+\cdots 3+2+1$. *The total is 4950 pairs of events for a single person in a single day.*

(Marks 2000, p. 237)

What does this mean? Consider the earlier example of your friend Craig. In this case two events – your thoughts on the bus and a chance meeting – turned out to have an astonishing similarity to each other: you thought about Craig, and there he was. In the same way, Jung's story about the patient's dream and the arrival of the beetle at the window is a case of two of these (roughly) 100 events per day resembling each other.

Now we can build on Marks and Kammann's insight. Assume there are about a hundred events per day that we might remember – especially if they were paired with a strikingly similar event the same day. So if we are impressed by an event with a probability of one in five thousand, there are more than five thousand opportunities for this to happen each day. In short, we might expect these 'astonishing' events to be happening all the time.

We can go further. French scientists and skeptics Charpak and Broch (2004), argue that in large populations – such as the tens or hundreds of millions in a modern nation – there will be a strong tendency for even less likely events to occur. Thus, if there is efficient reporting of unlikely events then, over a few years, we might find that we are looking at reports of coincidences the probability of which is one in many billions. This, one suspects, is exactly what happened in the case of Mr Page and Mr Pape above. That was the result of a national competition, so it is hardly surprising that the results seem very unlikely.

This argument is completely at odds with our strong feelings that these coincidences are amazing. There are several possible explanations why we might feel that the coincidences are far less likely to occur than they actually do. A psychological explanation – with some evidence – comes from Blackmore and Troscianko (1985). This essentially argues that our perceptions of probabilities are grossly skewed, so that we see unlikely events as being far more rare than they actually are. Other researchers (e.g. Kahneman and Tversky 1979) have produced evidence supporting this view, so it is not simply speculation. It is another case of the way our minds work creating the likelihood of paranormal belief. Psychologists continue to examine this aspect of the human mind (e.g. Brugger & Taylor 2003; Brugger, Landis & Regard 1990).

Clearly, there is no precise or exact way of working out what is truly amazing and what is not. Two statisticians (Diaconis & Mosteller 1989) examined this in some detail. The following points seem quite clear. First, coincidences that appear to be utterly astonishing at first sight may be a good deal less so when we consider how many coincidences could occur, but actually do not. Second, from the analysis of coincidences, it seems clear that the burden of proof is well and truly on those who claim that some principle other than pure chance is at work. Indeed, if we wield Occam's razor, we can see that since natural explanations exist, there seems to be no need for some new universal principle to explain what is happening. Finally, Sagan's balance requires an extraordinary level of evidence to substantiate the claim. Marks and Kammann's analysis does seem to make coincidences less supportive of extravagant claims about the nature of the universe.

Perhaps we should leave the analysis of coincidences there. It is an intriguing field, with much written about it. However, the main aim of this section was to show how intermediate concepts can enable the skeptic to make sense of apparently paranormal claims. In the process, we also saw that one of the most important phenomena, recruiting many people to paranormal belief, is much less convincing once it has been closely looked at.

CHECKING THE EVIDENCE – IS IT RELIABLE?

Scientists, as we have seen, are aware that individual scientists could be mistaken in their theorising and in their experiments. As a consequence, it follows that when scientists use results and theories put forward by other people, they scrupulously state where these results and theories are to be found. Of course, this has the effect of enabling scientists to check the reliability of other people's work: if they cannot use other published work correctly, doubts must be raised about anything else they say.

By the same token, it is always reasonable to check the reliability of paranormal claims. Frequently, books on these topics consist of amazing statement after amazing statement, often with little or no indication of where the information came from. In other cases, the books are full of references and reading lists, and the sheer quantity of evidence makes it hard to know what to do. Logically, if you want to know how reliable the claims are, you have to check some of them. But how can this be done without taking up enormous amounts of time?

James Trefil (1978) has suggested a simple method, which takes little time. That is to pick a few points and check those in depth. This gives a good quick indicator of whether the book as a whole can be believed. Of course, things can go wrong. You might pick the only reliable point in a book of rubbish, or the only bad point in a reliable writer's book. Still, if you check a few, this method will on average give you a good idea of what is happening.

In one case, I had the task of reviewing an enormous book, over 500 pages, by a well-known ufologist Timothy Good (1987). The book had a calm, thoughtful style, and its two key propositions emerged slowly as I read. First, there appeared to be an overwhelming mass of evidence that UFOs were being seen all over the planet: the book reported literally hundreds of cases in astonishing detail. Second, there seemed to be a worldwide cover-up of all this evidence by governments and secret services. The mass of evidence cited by the author was impressive.

I wanted to do a review of this book for a magazine. However, to check all the factual claims in the book would take years. What on earth was I to do? What I did was follow Trefil's advice. I picked some points and checked them.

Some of the evidence was feeble, even without checking. For example, Good quoted a case of a couple of electricity cables falling in South West England. The sole link to UFOs is that Good's press clippings at that time were only from the south of England. This is so weak that it hardly counts as evidence at all.

Other evidence looks astonishing at first glance. For instance, test pilot Robert White is reported as exclaiming in flight 'There are things out there. There absolutely is'. This certainly sounds impressive. However, Good omits to describe what White was talking about. So I had to go to the reference (*Time* magazine, 27 July 1962). This revealed that White was talking about an object looking 'like a piece of paper the size of my hand'. This is hardly strong evidence for UFOs, and it suggests that Good's evidence is nowhere near as strong as he suggests.

In addition, there is another way of checking out Good's claims. American skeptic Philip Klass has investigated many UFO claims (e.g. Klass 1974; 1976; 1983). Klass's books came out years before Good's, and Good has corresponded with him. Klass has acquired a reputation as a fearsome debunker of UFO claims. It follows that we can learn much from the way that Good treats Klass's objections to the ufologists' claims. If Good produces strong reasons for discounting Klass's objections, that is a clear mark in his favour. However, if he neglects the problems raised by Klass, or treats them inadequately, then that is an indication that the use of evidence in the book is not reliable.

Let's have a look at an example. What about UFOs over Iran in 1976? This is a highly dramatic story, to which Good devotes about three pages. Stripped of details, a couple of F4 jets from the Iranian airforce were scrambled to investigate some sightings of lights in the sky. Reports vary, but at least one jet suffered grave electrical failures, tried to fire a missile at something and had something fired at it. A nearby jet airliner also suffered radio failure. Good spends most of his three pages reproducing some reports by a US Defence Attache, and then quotes an evaluation of the report as a 'classic'. That's pretty well all he does.

What about Klass? Years before Good published his book, Klass made repeated attempts to find out more. He wrote to a number of people involved, often receiving no reply, and found a sizeable number of pieces of evidence that were not contained in the original report. First, Klass found that only one aircraft had suffered electrical malfunctions, not two. What is more, that plane had a history of unexplained electrical faults, and the electrical workshop responsible for it was notorious for poor performance. In this context, a temporary electrical malfunction can hardly be characterised as mysterious. He also points out that the aircrews at the time were tired and rattled, and could have mistaken stars or meteors for UFOs and 'missiles'. In addition, Klass points out that radio faults on airliners are not unknown, and that is why they carry backup radio sets.

I worked through a number of Good's claims in this way, and kept finding that the evidence, while presented as being of good quality, was actually very feeble. If this is true of the whole book, then it has very little value, being simply a mountain of bad evidence. In short, I emerged profoundly unconvinced by Good's argument and evidence.

Have I shown that there is nothing of value in Good's book? No, it might be that, purely by chance, I have picked out all the bad pieces of evidence in an otherwise excellent work. Or, Good's book might indeed be mostly poor quality, but hidden somewhere in its pages, there might be a single conclusive case which would convince almost anyone of Good's theories. On the other hand, I have done some checking, and in my view the burden of proof is on Good. If he wants to establish his theory, and convince others of it, then it is incumbent upon him to produce the evidence. On this showing, so far he has not. Weighed in Sagan's balance, the evidence does not seem convincing enough for the amazing claims that are being made.

THE RELIABILITY OF REFERENCES TO SCIENCE

The Good case is not by any means the worst example of bad evidence I have encountered. Many others are far worse. One of the most shocking cases of unreliability happened to me, long before I became an active skeptic. I had developed a small talk on Erich von Däniken which I gave to assorted groups in schools, small societies and the like. I also turned it into a small booklet (Bridgstock 1978). In essence, I carried out similar checks on von

Däniken as those on Goode, and I concluded that Däniken's evidence was pretty well worthless. Other researchers have come to the same conclusion, such as Castle and Thiering (1973), Story (1978) and Wilson (1972).

At one time I was talking to a group of science fiction fans. I gave my talk on Erich von Däniken, and indicated at the end that there were other beliefs that were as badly supported as von Däniken's. I mentioned a few examples, which included creation science. And there was a creationist supporter in the audience. He approached me afterwards, as asked if I had ever studied any creationist literature. As a matter of fact, I hadn't. Well, he asked, would I be prepared to do so? I was trapped. Of course I would. What else could I say? And so I found myself looking at a handful of highly coloured creationist magazines. They seemed to tell a devastating story. Apparently the theory of evolution was about to be swept away by the new creationist theory. This was the theory spelt out in the book of Genesis. The creationists were quite frank about their religious commitment, but they also used scientific evidence. Prominent scientists were quoted as doubting evolution, and many examples of creationist evidence were cited. Throughout the magazines, reputable scientific journals such as *Science* and *Nature* were referred to. The creationists seemed to have a good case.

What was I to do? It took me weeks to do the obvious. I took the magazines to the university library and checked some references. I knew that there were some errors in virtually all scientific work, but if the creationists were even mostly right, they seemed to have much evidence on their side.

It took several weeks to check a reasonable number of references, but the picture became clear quite quickly. The creationists' references were hugely, overwhelmingly wrong. After a few weeks' work, I arrived at two generalisations about creationist referencing to science. First, about 90 per cent of all creationist references contained some major error. This would be a misquote or a misrepresentation so gross that it completely invalidated the creationist claim. In addition, on average, each creationist reference to science had a major error, and also a minor one, such as a wrong page, date or author (Bridgstock 1986a). An example of a major claim may make the point clearer. One of the first references I checked was by the distinguished British scientist, Lord Zuckerman. In a creationist publication *The Quote Book* he was quoted as saying (and the words were in quotes): 'If man evolved from the apes he did so without leaving any trace in the fossil record.' (Mackay et al 1984, p. 11)

The quote apparently came from Lord Zuckerman's book *Beyond the Ivory Tower* (Zuckerman 1970). The implication is clear: this eminent scientist has grave doubts about the evidence for evolution provided by the fossil record. But is this true? I checked the quote, and Lord Zuckerman actually wrote:

> For example, no scientist could logically dispute the proposition that man evolved from some ape-like creature in a very short period of time, speaking in geological terms, without leaving any trace in the fossil record.
>
> (Zuckerman 1970, p. 64)

This is a quite staggering misquote, and it is hard to believe that anyone could seriously present the first quote as being the same as the second. Yet there it was, in the *Book of Quotes*, referenced to the appropriate page. Such misrepresentations are, unfortunately, very common in creation science. Another type of misrepresentation is when the actual reference to the evidence is correct, but it is torn out of its context so that the meaning is changed. One example might make the matter clear. Henry Morris, the founder of the modern creationist movement, wrote a textbook *Scientific Creationism* (Morris 1984), which tries to summarise a great deal of evidence for the creationist viewpoint. One point he makes is that according to two oceanographers, Riley and Skirrow, the input of elements into the oceans would simply not have taken billions of years to reach their current levels. Here is part of Morris's table that summarises his evidence.

Chemical Element	Years to Accumulate in Ocean from River Inflow
Sodium	260,000,000
Magnesium	45,000,000
Silicon	8,000
Potassium	11,000,000
Copper	50,000
Gold	560,000
Silver	2,100,000
Mercury	42,000
Lead	2,000
Tin	1000,000
Nickel	18,000
Uranium	500,000

(Morris 1984, p. 154)

On the face of it, Morris has a good question to ask. If the Earth is indeed billions of years old, why are the concentrations of elements in the oceans so low? Does this not indicate that claims that the oceans are billions of years old are, at least, rather shaky? Although the apparent ages of the oceans do vary from metal to metal, it does look as if Morris has made a good point.

To answer Morris's question, I made my way to the Geology Library at the University of Queensland, and did some checking. It was not hard to find Riley and Skirrow's (1965) book *Chemical Oceanography*. The first surprise came when I looked at the pages indicated by Morris. In fact, the author of the article is not Riley and Skirrow (they were the book's editors) but a scientist named Goldberg. The enormity of this is hard to exaggerate: Morris didn't even get the *author* right! Then, it was not difficult to locate the table used by Morris, and here is part of it:

Abundance of the elements in sea water and residence time		
Elements	Concentration in mg/l	Residence Time (years)
:		
Na	10,500	2.6×10^8
Mg	1,350	4.5×10^7
:	:	:
Si	3.0	8.0×10^3

(Goldberg 1965, p. 164)

The notation is different, but the elements and numbers are identical. Na is the chemical symbol for sodium, Mg for magnesium and Si for silicon. There are many more elements in the table, but we will focus on these three. For each element, the numbers in the right-hand column are the same as the corresponding one in Morris's table. I had found Morris's source of evidence. However, a look at the top of the table shows that the relevant column is not 'Number of years needed to accumulate in the oceans'. Instead, it is something called 'Residence Time'. If 'Residence Time' means the number of years needed to accumulate in the ocean, then Morris is justified in renaming the numbers in this column. If 'Residence Time' means something else, then he is misrepresenting the evidence. In the text of Goldberg's paper, we find residence time defined in this way:

The residence time, τ, can then be defined as the average time which it remains in the sea water before removal by some precipitation process.

(Goldberg 1965, p. 172)

This is completely different from the way that Morris has used the numbers. The length of time that an element remains in the water has absolutely nothing to do with the time it has taken to accumulate to the current level. Although the numbers are correctly stated, their meaning has been misrepresented. This is a good example of ripping something from context, so that it says what the creationist wishes it to mean, rather than what it actually means.

The point that arises from these two examples is pretty simple: know the reliability of your evidence. Make a few checks, so that you know how good the evidence actually is. In general – and with exceptions – the use that scientists make of other scientists' evidence can be assumed to be fair. On the other hand, if you are investigating a paranormal claim, it is foolish to begin by taking the evidence quoted at face value. At the very least, a few key pieces of evidence should be checked, to see how reliable the claimant is. With the creation scientists, it soon became clear that their work could not be relied upon, and had to be checked in great detail. So Good, von Däniken and the creationists all fall into the same category: their evidence is simply not reliable.

This is a terribly important point. We humans are not very reliable reporters, even of evidence which is in front of us, especially when deep passions and beliefs are involved. We should therefore always be extremely careful when we look at reported evidence. It may simply not be reliable enough to take seriously.

THE DOUBLE-BLIND RANDOMISED CONTROLLED TRIAL

There are other ways in which sincerely reported evidence can be completely wrong. Imagine a friend of yours has been suffering from an illness. You do not see your friend for a few weeks, then one day you happen to meet up. And your friend is well. 'Yes', she tells you, 'I went to a naturopath and he spent an hour talking to me. He gave me some preparations and I felt better straight away.'

Such evidence is among the most powerful we can experience. It comes, often, from someone we can trust, and someone with no interest in selling us anything. And it comes from direct personal experience. So does this not mean that we should accept our friend's evidence? Probably not, in the view of most skeptics. In assessing the effectiveness of alternative medicines, the most convincing type of evidence is known as the randomised double-blind controlled trial. We will look at what this consists of and see why it is so important.

In a randomised double-blind controlled trial, patients are assigned at random to one of two groups. One group – and neither doctors nor patients know which – is given the drug that is being tested, the other, usually, a drug that does nothing at all. When the patients have been taking the drug, or the useless preparation, for a suitable length of time, they are examined by doctors. Of course the doctors keep records of the patients who have been cured, of those who have improved, and so on. Only at the end, when all the data collection is complete, is it revealed which patients received which drug.

There are variations on the randomised double-blind controlled trial. For example, the group who are not receiving the new treatment may not receive a worthless preparation, but might be given an established drug the effects of which are well established. This enables the researchers to see whether the new drug is as good as the existing medicine, or better.

We can now understand all the terms in the title: randomised double-blind controlled trial. The patients are assigned at random to the treatment groups. Neither the patients nor the doctors know who has received which treatment until after the test is over. And while there is an experimental group, receiving the new treatment, there is also a control group, receiving a known treatment, or an ineffectual treatment, to which the effects of the new treatment can be compared.

Now the obvious question: Why bother with all this complexity? Why not simply try new medicines on patients and see if they work or not? What is the need for the double-blind controlled trial? The answer is that without the double-blind controlled trial any one of a number of effects can grossly mislead us about the effectiveness of medicines. Perhaps the best known and most important of these misleading effects is the placebo effect.

THE PLACEBO EFFECT

Imagine you are ill. Perhaps you have a slight fever, pains, feelings of nausea or exhaustion. You go to see a naturopath, or a homeopath, or perhaps an acupuncturist. The therapist gives you a careful hearing, and prescribes some treatment. Lo and behold, after the treatment you feel distinctly better. Have you not just proved that this alternative treatment actually works?

Well, not necessarily. This type of evidence is called 'anecdotal'. Often an experience of this type can convince a patient that a course of treatment has done something for them. What is more, if the patient goes out and tells other people, they may give it a try as well. After all, what is more convincing that something one has experienced oneself?

From the other side of the desk, alternative medicines can seem equally valuable. Consider, for example, Dr Steven Bratman, a perfectly genuine physician who is registered to practise in the American state of Colorado. Dr Bratman stands quite high in his profession:

> Dr. Bratman has been an expert consultant to the State of Washington Medical Board, the Colorado Board of Medical Examiners, and the Texas State Board of Medical Examiners, evaluating disciplinary cases involving alternative medicine. In addition, he has performed medical review of the advertisements associated with natural products.
>
> (Bratman 2006)

What is Bratman's experience of alternative medicines? Here is what he has to say of the time when he used them:

> During my years of practice, many of my patients reported benefits, and I could often see those benefits with my own eyes. Use of medications declined, days off work decreased, objective measure of urine flow and liver enzymes and blood pressure shifted toward the normal range, and injured joints recovered normal range of motion.
>
> (Bratman 2005, p. 64)

Both the patients' reports of feeling better, and the physician's report of improvements seem to be powerful evidence. Even by skeptical standards, are we not justified in concluding that some alternative medicines, at least, have beneficial effects?

Not necessarily. Immediately after the passage quoted above, Bratman goes on to say this:

> I am now quite convinced that my eyes were wrong, and most of the treatments I provided were nothing more than elaborate placebos.
>
> (Bratman 2005, p. 64)

This is an astonishing statement, and it probably requires a good deal of moral courage to make it. As a teacher, I know there is nothing more gratifying than seeing students learn. In the same way, presumably, there is nothing better for a devoted doctor than seeing patients improve under their care. Yet Bratman appears to be saying that he had little or nothing to do with the patients' improvement. Why?

There are several answers to this question. Perhaps the most important one is the 'placebo effect'. Imagine that you are a doctor, and you are excited about a new drug, which promises to be able to cure a serious illness. You administer the drug to a group of patients suffering from the disease, and many of them report feeling better. Can you conclude that the drug is effective? Alas no, because even if the drug were useless, a percentage of people would report that they felt better anyway.

The placebo effect can be quite substantial. Beecher (1955) reported that 30 per cent respond to a placebo, saying that it reduces pain. More recently, drawing upon more advanced studies, Hoffman, Harrington and Fields (2005, p. 255) cite studies showing that the proportion feeling less pain can vary from 27 per cent to 56 per cent. The placebo effect is very widespread.

This is the kind of effect that Francis Bacon was talking about earlier. However, he was referring to the negative case where a dire prediction could trigger an illness of some kind. Here, we are talking about a good prediction: a placebo is in effect a promise that if you take the drug you will improve. Yet there is no real improvement, in the sense that the drug has no genuine therapeutic effect.

What can be done? An obvious answer is to give half the patients the new drug, and the other half a placebo – a useless medicine – and then see whether the real drug works better than the useless one. This is a controlled trial, of the very simplest kind.

The problem with this approach is that the doctor knows which patients are taking the real drug, and this may affect the patients' reactions – or the doctor's assessment of the results. The only way to counter this problem is to arrange matters so that neither the doctors nor the patients know who is getting the real drug, and who is getting the placebo. This leads us to the double-blind controlled trial. If we then randomise the patients, so that the ones in the treatment group and the ones getting the placebo are very similar, we have the best form of evidence for the value of new medicines.

This is why Steven Bratman was so devastated when he learned of the importance of this type of research. All of his experience of patients improving was ruled out at one stroke: in the absence of double-blind controlled trials, stories of patients getting better simply carry no weight at all. No wonder Bratman described this realisation as being like a tornado going through his mind.

Bratman deserves a good deal of credit for his courage. There are numerous ways he could have evaded the issue, and somehow managed to preserve some kind of pretence that alternative medicines have some validity. Instead, he came forward and clearly spelt out the issues as he saw them. He saw that the placebo effect is not the only way in which a completely useless treatment can appear to be of value. He lists several more such effects, and any reasonable skeptic must know about them too.

Two such effects are closely linked to the placebo effect. One is called observer bias, the other the study effect or the Hawthorne effect. In the first case, a doctor administering a treatment in which she believes, is far more likely to see benefits in a group receiving the treatment than in a placebo group. Thus, both patients and doctors are likely to be influenced by the belief that a new and effective treatment is being given. This is the key reason why the double-blind controlled trial is so important. The study effect is simply the observed fact that people who know they are being observed, or being used as part of a study, often 'rise to the occasion', that is, they behave in ways quite different from normal. They may act in ways that minimise the symptoms of a disease, or may, if at work, be much more productive. Both the effects of observer bias and the study effect can ruin evidence about the value of medical treatments, and so the double-blind controlled trial must be used to eliminate the problems (Bratman 2005).

Bratman does not stop here, though. He goes on to outline two more influences which can also lead people to believe that a treatment has value, when in fact it is worthless. One is called 'natural course of the illness', the other is the 'regression to the mean'. The first effect is very simple. Most illnesses are self-limiting. If we have the flu, we suffer for a few days, and then we get better. If we feel depressed, we usually feel happier a day or two later, and so on. Thus, if we are feeling ill, there is a good chance that we will soon start to feel better. Our error is not in noting that we feel better, our error is in attributing it to the treatment, when in fact it is the recuperative power of our own body that is doing the job.

Regression to the mean is more complex. It is a statistical effect which need not have anything to do with medicine. It arises from a very simple proposition: that where we have been lucky (or unlucky) once, we will probably not be as lucky (or unlucky) the next time. This is so obvious and simple that it seems to need no explanation, but it gives rise to a whole set of complex illusions.

As a simple example, let us imagine that we invite hundreds of people to throw a couple of dice and add up their scores. The people with the very highest scores will, they are told, go on to the second round of the competition, with a prize for the highest scores to be awarded at that point.

After the first round a small proportion of people – about one in 36 – will have thrown the maximum score of 12 (two sixes). They go forward to the second round and throw again. And, of course, only a few of them will throw a score as high as in the first round. From simple statistics, only about one in 36 of these people will throw the same score again. The scores of the rest will have 'regressed' towards the mean. Of course, something similar would have happened if we had invited the dice-throwers with the lowest totals to throw again: their scores would have risen.

Clearly, regression to the mean can also affect our medical condition. Our temperatures, blood pressure and feelings of wellbeing can all vary without rhyme or reason. Thus if we take a medicine when we are feeling bad, there is a real chance that we will begin to feel better regardless of the effects of the medicine: it is the regression towards the mean which is having the effect.

All of these effects – placebo, observer bias, study effect, natural course of illness and regression to the mean – can make a medicine or treatment seem effective when in fact it is not. Only a double-blind controlled trial can reveal whether a treatment really has valuable benefits, or whether one of these effects is misleading us. Thus anecdotes from patients, and multiple stories from doctors, should have no effect on our judgement of the validity of treatments. Only the evidence from double-blind controlled trials should be acceptable.

There is worse to come. Even the results of the most immaculately conducted double-blind trial may be misleading as well. This is because of the nature of statistical testing. Without going into detail, the results of a double-blind trial will be assessed using statistics. That is, calculations will be done to see how likely it is that the results could have been the result of pure chance. By convention, results need to have less than one chance in 20 (or five per cent) for them to be regarded as significant. Some experimenters take a tougher line, and demand that the results should have less than one chance in 100 (or one per cent).

The problem should now be clear. If a substantial number of tests of the usefulness of some procedure are done, a small fraction of them, purely by chance, will meet the level of required significance. Hence, dotted through the literature one is likely to find reports of this kind. And it is very easy, by selectively quoting these reports, to make a powerful-looking case for the efficacy of a treatment which is in fact useless.

In fact, it is worse than that. There is a bias in the research literature towards studies which yield statistically significant results. Hence, the research reported in the literature is not representative of all research. It might therefore be that a scrupulously careful search of the literature, taking account of every relevant study in print, might conclude that a medical treatment had value when in fact it does not. This has been termed the 'file-drawer' problem (Kennedy 2004). The name comes from the observation that negative studies – without statistically significant outcomes – are likely to end up in the file-drawer rather than in print.

In this section we have dealt at some length with an important set of skeptical concepts. They show why worthless remedies can sometimes appear to be extremely beneficial. There is not one but a whole series of effects which can produce these bogus results. The best way of controlling for these misleading claims is to insist that evidence must come from double-blind controlled trials, and even then, the results can occasionally be misleading. Still, insistence upon double-blind trials is at least a first step and will eliminate many false claims on its own.

HOAXING, FAKING AND SELF-DECEPTION

The placebo effect, and the other illusions, gives us several ways in which apparently paranormal results can be achieved. A faith healer or alternative therapist can make people feel better, even if there is no actual long-term benefit, and it takes something like a double-blind controlled trial to weed out what actually does work from what does not.

So far, we have assumed that the people making the paranormal claims are sincere, and that they truly believe in what they are saying. In fact, this is not always the case. We all know that human beings are imperfect, and that we vary widely in honesty, as in most other features. Some of us are psychopaths, with no conscience or concern for the feelings of others. And at times almost all of us lie, or deceive, or fall short of our own standards. It is part of being human. The same applies in the paranormal area. Since people concerned with the paranormal are human, we might reasonably expect them to show the full range of human virtues and vices. We might expect to meet some ruthless exploiters, some small-time con artists, some self-deceivers, some people on the verge of madness, and so on. Indeed, it would be truly astonishing if we did not meet such people.

Earlier in this chapter we encountered some people whose work was grossly misleading. The creation scientists misquoted and misrepresented scientific work so that it seemed to support their own position. However, from a few encounters with them, I did not form the view that they were psychopaths or conmen. My view was that their belief in their – narrowly fundamentalist – religious position was so strong that nothing else mattered by comparison. Driven by their beliefs, I suspect, they regarded the evidence as being almost unimportant by comparison. The important thing was to win souls, to bring people to awareness of their God almost regardless of the accuracy of the evidence they used.

It is easy to condemn the creationists for their attitude, but the impression remains that these are not cold-blooded frauds. They are seeking to impart their understanding of the purpose of life, and to do so, are prepared to flout what they regard as less important rules. They can be condemned, but they should not be regarded as totally insincere.

It seems likely that many paranormal practitioners fall into this intermediate area. They are guilty of sometimes committing fraudulent acts, but they are 'sincere' in that they believe strongly in their basic thesis. Perhaps they rationalise that although the paranormal is real, sometimes the results have to be faked in order to lead other people to the truth as well. My suspicion is that a good deal of apparent faking in the paranormal stems from concerns of this kind. Of course, it means that we should always bear in mind that other people have different standards of truth from ourselves, and so check the evidence.

Of course, the paranormal also has its share of outright fakes. In the Introduction we saw the case of the young woman in Texas who was conned out of thousands of dollars by a 'clairvoyant' who was eventually arrested (Davis 2005). M Lamar Keene, a leading medium, has told of the cynical manipulation and exploitation that accompanies gullible clients. He writes 'The facts are that from its very beginning modern spiritualism has been riddled with fakery, humbug and deceit.' (Keene 1997, p. 115) Even where the psychic believes they have genuine powers, there may still be manipulation to increase payments (e.g. Woodruff 1998). The moral is simple: check the evidence, and do not assume that other people see truth the same way as you do.

This brief review of modes of dishonesty in the paranormal shows how careful one should be. Before accepting any paranormal claim, the evidence should be checked most carefully, as the results can be disastrous.

LAUNCHING PAD

This chapter has addressed the existence of a 'missing link' in skeptical thinking. The large-scale principles of skepticism are useful, but they are usually far too abstract. For each type of paranormal claim made there is also a set of intermediate ideas which help the skeptic understand what is going on. We have looked at some of the most useful ones here. There was the astonishing power of coincidences to explain apparently paranormal events which we see happening. There is the amazing capacity of information to 'leak' through apparently impermeable barriers and in that way confuse the most carefully constructed experiments. Then there is the power of a whole range of effects, some of which are the placebo effect, regression to the mean and the study effect, which make it appear as if medical treatments are effective when in fact they are not. Then we investigated the reliability of eyewitness claims and of human memory. Although these may not be totally misleading, it is clear that they are not as strong as their proponents would like to believe. Finally, we looked at the reliability of paranormalist evidence in the context of the claims of creation scientists. It is clear that the evidence cited by these people is grossly misleading and cannot be taken seriously. In general, it looks as if the paranormal has its share of crooks, deceivers and self-deceivers, which may render any evidence valueless.

Where does this lead us? First, it tells us that merely because someone claims that there is evidence for some paranormal claim, it does not follow that there is no natural explanation. As we have seen, some of the evidence for paranormal claims is extremely poor, and in other cases, closer scrutiny has shown that natural explanations do exist.

Second, merely because there are no obvious explanations, and no obvious ways of bringing the skeptical principles to bear, it does not follow that skeptical explanations do not exist. It took Marks and Kammann years to find the fatal flaw in the remote viewing experiments, and it took weeks of work for this author to realise the truth about creationist 'evidence' for their beliefs. (In short, skepticism can involve a great deal of work!)

Third, it also seems to follow that one of the most important abilities for a skeptic to have is the ability to find out what work has been done before. Amazing coincidences look a great deal less amazing after reading Marks and Kammann's chapter. In almost every area of paranormality, there exists a substantial body of skeptical work that can be used to throw light on amazing claims.

We now have a grasp of the principles of skepticism, and how they can be brought down to earth via the intermediate concepts. In the Chapter seven, we will look at the question of whether skeptical principles can be applied to non-paranormal phenomena. As we shall see, this has the potential to broaden the importance of skepticism a great deal, and indeed to make it a key feature of modern civilisation.

Here we have dealt with the awkward interface between skeptical ways of thinking and the claims made for the paranormal. As we have seen, there is an enormous mass of evidence which can explain most, if not all, paranormal claims. The difficulty is to locate this material and use it. Kelly (2004) is essentially a mass of intermediate concepts. In the introduction, Kelly shows that she is aware of the skeptical approach, but almost all the book is devoted to showing how specific ideas and findings can explain claims of the paranormal naturally. An equally useful way of doing this is to access *The Skeptics' Dictionary* (Carroll 2008) on the internet, and look up the phenomena which interest you. Or simply browse, to gain some idea of what skeptics have discovered.

Anyone with an interest in the paranormal should understand the importance of the placebo effect and other related processes. The paper by Bratman (2005) is a good overview. In addition, almost anything by Elizabeth Loftus is good at showing how fallible human memory can be (there are several of her works in the bibliography). Clancy's (2005) recent work is a detailed study of why people believe in alien abduction. It is vivid and fascinating. Keene (1997) is hard to obtain, but is a searing exposure of what some mediums truly think of their clients.

Marks (2000) gives important understandings of several paranormal fields. I especially liked the chapter on coincidences. For people with a statistical bent, Diaconis and Mosteller's paper (1989) shows how professional statisticians look at the problems raised by these amazing events.

6 | Skepticism, ethics and survival

W E HAVE SEEN that paranormal belief can have extremely dire effects. The deaths of Caleb Moorhead and Liam Williams-Holloway were both, in part at least, probably caused by the paranormal beliefs of their parents. We also saw that a young woman's credulous approach to paranormal beliefs made her vulnerable to a 'clairvoyant' who relieved her of large amounts of money. We have also seen that many skeptics have concerns about the suffering caused by some paranormal claims, and the threat they pose to science.

This suggests an important dimension to scepticism, an ethical perspective, and we will look at this more closely in this chapter. The implication is that in some cases skepticism may not simply be a useful set of intellectual tools, but an approach to human affairs that can have profoundly beneficial effects. If this is so, then we should be able to discern some concern within skepticism for right and wrong conduct. That is, skepticism has an ethical dimension, giving us guidance about right and wrong conduct. We will begin by asking a rather surprising question: can it be unethical to hold certain beliefs? As we will see, some very serious thinkers argue that this is so. We will also look at some aspects of skeptical behaviour and will try to sketch an outline of a skeptical ethic.

A major approach to ethics is termed 'consequentialism' and it focuses on the consequences – the results – of particular courses of action. Actions are evaluated according to their overall effects for or against human happiness. Gradually, consequentialism is becoming the predominant approach in ethical theory. As we will see, it also yields valuable guidance within skepticism. We will adopt a consequentialist approach to ethics in skepticism and try to use it to solve some particular problems that skeptics may face. There are other approaches to ethics – notably deontological approaches and virtue ethics – but the consequentialist approach yields useful results very quickly.

More generally, the question has to be faced, why be skeptical at all? If there are personal benefits in believing for the sake of believing, why bother with the rigours of the skeptical approach? We set out to answer that question in this chapter. We will see that there are broader philosophies into which skepticism fits very comfortably, and which strongly support the use of a skeptical outlook.

CLIFFORD AND THE ETHICS OF BELIEF

We will start by looking at one of the most important, and least-known, skeptics, the 19th century British mathematician William Kingdon Clifford. Clifford was only 34 when he died, but he had lived an amazing life. He had become a professor of mathematics, survived a shipwreck in the Mediterranean – and also presented an important paper to the Metaphysical Society in London (Clifford 1879). This paper argued that it is actually unethical to believe in certain circumstances.

What is Clifford's argument? He begins with a striking example. Imagine the owner of a ship, which was about to carry emigrants to a new land: 'He knew that she was old, and not overwell built at first; that she had seen many seas and climes and often had needed repairs.' (Clifford 1879, p. 177) Quite reasonably, the shipowner has doubts about the safety of his ship. However, he puts aside all thoughts of inspections and repairs, as these would cost a great deal of money. Instead, he convinces himself that the ship is safe, and that all will be well. Then there is one terrible sentence 'and he got his insurance-money when she went down in mid-ocean and told no tales.' (Clifford 1879, p. 178)

Of course, the shipowner behaved very badly, but Clifford's argument does not stop there. Not only was the owner guilty of the death of those people, but he had done something else wrong:

> It is admitted that he did sincerely believe in the soundness of his ship, but the sincerity of his conviction can in no wise help him, because he had no right to believe on such evidence as was before him.
>
> (Clifford 1879, p. 178)

Then Clifford turns the argument around. What if the ship had made its trip safely? The people all reached their destination without incident. Was the shipowner thereby cleared of any liability?

> Will that diminish the guilt of the owner? Not one jot. When an action is
> once done, it is right or wrong for ever: no accidental failure of its good or
> evil fruits can possibly alter that. The man would not have been innocent,
> he would only have been not found out.
>
> (Clifford 1879, p. 178)

So for Clifford the key to ethics and belief is that we do not have the right
to believe without adequate evidence. The fact that a belief is true or false
does not affect the ethics involved in holding it.

Clifford gives another example, about agitators persecuting a minority religious community without sufficient evidence (Clifford 1879, pp.
178–9). The unjustified attacks cause great misery before being shown to
be unfounded. The point is the same: holding beliefs without adequate
evidence, whether the beliefs turn out to be true or not, is unethical. We
have no right to hold beliefs on insufficient evidence.

On the face of it, Clifford's argument is counter to some of our most
cherished notions of liberty. Don't we accept that people have a right to
their own opinion, whether it is religious, political or paranormal? Have
not many dictators – Hitler, Stalin and Mao – persecuted and murdered
people who disagreed with them? In short, is not Clifford's argument an
invitation to totalitarianism?

Let us go a little deeper into Clifford's argument. Why does he believe
that it is unethical to hold some beliefs in certain circumstances? Essentially,
he has two arguments. The first is obvious from the examples. Sometimes,
holding unwarranted beliefs can lead to terrible consequences. A shipload
of people perished in the ocean because of the unjustified belief of the
owner. The lives of people in a small community were made miserable in
the second example, because of unjustified beliefs held by the agitators. So
the beliefs were a major cause of the unhappy result.

Clifford's second argument is more subtle. He notes that the first argument can be countered. Regardless of a person's beliefs, it is the actions that
caused the bad consequences. If the shipowner, regardless of his beliefs,
had had the ship checked, all would have been well. If the agitators had
examined some evidence about the minority, they would have realised that
the persecution was not justified. So why does Clifford condemn the beliefs
and not the actions which led to the disaster? Clifford has an answer to
this, and it probes deeply into our nature.

> No real belief, however trifling and fragmentary it may seem, is very truly
> insignificant; it prepares us to receive more of its like, confirms those which
> resembled it before, and weakens others; and so gradually it lays a stealthy

train in our inmost thoughts, which may someday explode into overt action, and leave its stamp upon our character for ever.

(Clifford 1879, pp. 181–2)

This is an important point. Not only can unjustified beliefs – in conjunction with actions – lead to disastrous consequences, but each unjustified belief can have the effect of lowering our intellectual standards, of making us more and more credulous, more and more prone to accepting other unjustified beliefs. In the long run, suggests Clifford, this can lead us to some forms of barbarism. Logically too, if holding unfounded beliefs makes us more likely to cause disasters, and holding one such belief makes us more likely to hold others, then the likelihood of disasters caused by our beliefs clearly rises.

There is another possibility that Clifford does not mention. If we adopt an unsupported belief, then it may not make us more credulous. Instead, we may apply different evidential standards to different statements. Creation scientists appear to do this: they strongly doubt anything that seems to indicate the truth of 'evolution-science' but accept uncritically anything pointing to creation (Kitcher 1982). If we develop the habit of holding different evidential standards for different beliefs, we run the risk of being hypocrites – or at the least of being inconsistent.

This is the heart of Clifford's argument. Unjustified beliefs – in terms of the evidence we have available – make it more likely that there will be disastrous consequences. In addition, unjustified beliefs make it harder for us to weigh evidence properly, and may also make us more credulous in the future. It is a stern indictment of what we normally think of as being a private matter.

Before going on, we should note that Clifford, in his paper, did attack religious belief, arguing that, like other beliefs, these should only be held if there were adequate reasons for belief. Since this is not relevant for us, we will not pursue this further, except to note that much of William James's defence, which we will look at next, was focused on the defence of religion.

WILLIAM JAMES'S REBUTTAL

What do we make of Clifford's argument? The first thing to note is that for about a century, the argument was neglected. This is because the great American psychologist William James (1950 [1897]) fiercely attacked Clifford's argument, and was widely thought to have refuted it.

James's paper is subtle, perceptive – and exasperating. He essentially treats Clifford's view as ludicrous, and hops from point to point without making a systematic argument. Perhaps three points emerge, one simply wrong, the other two partly right. The incorrect point is that James argues that we have the right to believe whatever we like, at our own risk. Perhaps, but the shipowner was not believing at his own risk: he was believing at the risk of other people. Indeed, one of Clifford's points is that we never believe purely at our own risk, as our beliefs affect the rest of society for generations to come. James does not address this point.

James's second point is somewhat better. He insists that almost no investigation is done without some prior belief. He is right about this. Studies by sociologists (Mitroff & Fitzgerald 1977) in science, for example, show that the most highly regarded scientists are often passionately committed to particular theories. Thus, in this case at least, Clifford's general principle does not appear to hold.

However, science is not typical of human activity. In science, as many studies have shown (Hagstrom 1965; Ziman 1968), there is a community of seekers after truth. Ideas held by one scientist can be criticised by another, and it is the opinion of the whole community that decides what is regarded as true. Most human affairs are not like this: there was no one to dispute the shipowner's opinion, only those who died when he got it wrong. Therefore, and perhaps contrary to their public image, scientists can safely be less impartial and less cautious about their beliefs than the general public.

James's third point is perhaps the best of all. He argues that there are certain decisions – he calls them 'momentous' – which we simply have to make. What is more, there is no chance of our collecting enough evidence to make a justified decision as Clifford would view it. In its most dramatic form, these conditions apply to religious belief. James accepts that we can never have sufficient evidence for religious belief, but argues that we must make a decision of some kind – after all, not making a decision is simply agnosticism – and so we should make the decision which best suits us. However, James also believes that momentous decisions of this kind also apply in non-religious circumstances.

Can such situations exist? They certainly can, though James does not give a good example. Here is one which, I believe, is momentous. Imagine that you are waiting for a train, alone, on a small suburban railway station. In the distance, you see the train approaching. At that moment a small child rushes onto the platform and, without looking where she is going, falls forward onto the tracks. She lies there, stunned but clearly alive. At that moment the train begins to enter the station. The driver sees the child and jams on the brakes, but the train cannot stop in time. Unless you do something, the child will be killed.

What can you do? You are not sure of your own ability. Can you jump onto the track, grab the child and jump back in time to save the child? If you can, you will save the child's life. If you try and fail, both you and the child will be killed. You do not have enough evidence about your ability to help, and there is not enough time to collect more. You must simply, as James says, make the best decision you can with the knowledge you have. Clifford's ethics do not apply.

In his argument, James is mainly concerned to defend religious belief against the attacks of atheists and agnostics. There is no doubt that this attack was part of Clifford's argument. James was arguing that religious choice is a 'momentous decision', where we must make a decision in the absence of adequate evidence. We may freely grant James this point, and also that where momentous decisions are concerned, we must also make decisions on the basis of inadequate evidence. Still, this does not dispose of Clifford's argument. In other circumstances, where we could collect enough evidence to make a decision for or against a belief, is there not an ethical imperative to do so? James has perhaps limited the validity of Clifford's argument a little, but he has not touched its heart at all.

Clifford's argument can be criticised in other ways. There is the point that Clifford's idea of belief is a good deal less sophisticated than that of David Hume. For Clifford, apparently, belief is an all-or-nothing affair: either you believe or you do not. However, there are shades of belief. As David Hume (1980 [1748]) put it, a century before Clifford, 'A wise man proportions his belief to the evidence'. In other words, a person can believe tentatively, more strongly or with great assurance. This is not a minor point. The shipowner might very reasonably have said to himself, 'I am reasonably sure that my ship is seaworthy. However, since lives are at stake, I will have the ship inspected for safety.' The strength of our beliefs can have a strong bearing on our actions, and on their consequences. Clifford seems to have missed this altogether.

Another important point is that Clifford's ideas of right and wrong are also black and white. Most of us would agree that both murder and theft are wrong, but that murder is worse than theft. Clifford does not seem to realise that some actions are worse than others. This raises the possibility that in some circumstances we might opt to do something which is wrong, in order to avoid a greater wrong. An obvious example might be to torture somebody in an effort to find out where a nuclear device is, which otherwise will destroy a city.

We can go a little further, and suggest that some paranormal beliefs may actually be beneficial. They can sometimes give comfort and may sometimes lead to wise decisions. This suggests that, first, skeptics should not be fanatical killjoys, but should have a broadly tolerant approach. It also suggests a point we will make later in this chapter, that we need to sort out which paranormal beliefs are most dangerous, and to focus on those.

This leads to another important point. Clifford's entire argument fails if we accept that beliefs need not be immovable. If a belief can be discarded, and someone genuinely will discard it when the evidence so indicates, then the wrong appears much diminished. More generally, we can retain the advantages of both James's and Mitroff's arguments if we postulate that it is acceptable to hold beliefs provided there is a stronger commitment to discard them if the evidence shows they are wrong. This means that the 'wrongness' of belief without evidence is negated by this greater commitment.

Can Clifford's rule be followed? Not according to some commentators. Jack Meiland (1980), for instance, believes that if we follow Clifford, we must believe at will. However, that is not what Clifford wants. Clifford does indeed refer to the shipowner convincing himself that the ship is seaworthy. However, that is what Clifford opposes. What Clifford wants us to do is to refrain from believing when the evidence does not justify it (Zamulinski 2002). That is much easier, and falls within the range of our capacities. It also means that Clifford's argument is completely different from the demands of totalitarian dictators: none of them wanted their subjects to check their claims against evidence before believing them.

Of course, in our complex modern world we cannot know everything. It is simply beyond our capacity to investigate every question. Clifford has a partial answer for this. Rather than working through his argument, it will probably suffice to give his conclusions, which are as follows:

> We may believe what goes beyond our experience, only when it is inferred from that experience by the assumption that what we do not know is like what we know.

> We may believe the statement of another person, when there is reasonable ground for supposing that he knows the matter of which he speaks, and that he is speaking the truth so far as he knows it.

> It is wrong in all cases to believe on insufficient evidence; and where it is presumption to doubt and to investigate, there it is worse than presumption to believe.

> (Clifford 1879, pp. 210–1)

Put in this form, Clifford's arguments resemble modern skepticism. The difference is that skepticism assumes a commitment to investigate the paranormal, while Clifford's concern is to underpin the use of evidence with an ethical commitment. It seems fair to say that Clifford's arguments, while they have been limited by William James, and are grossly oversimplified, are by no means refuted, and can stand as the basis of an ethical commitment to skepticism.

However, Clifford's argument is much weakened by the point that the degree of ethical evil can vary, and we can avoid much of the harm if our beliefs are always subject to a willingness to discard them in favour of contrary evidence.

ETHICS AND THE SKEPTIC

This extension of ethics into belief has some rather unexpected consequences. We have seen that Clifford's argument is consequentialist in nature; believing in the absence of evidence is wrong because of the consequences it has. What consequences are these? There are two: first, that sometimes believing without evidence can lead to disasters (remember Clifford's unseaworthy ship). Second, Clifford argues that believing without evidence also corrupts our thinking, it makes us more credulous. And we have seen that credulity is not the only consequence. We might also, as another consequence of believing without evidence, acquire a split system of reasoning: things that suit us are believed without evidence, propositions which do not suit us are not believed. The likely consequence of this irrationality is an increase in human suffering.

From this viewpoint, skepticism – demanding evidence for paranormal beliefs – is not simply an enjoyable hobby. It is an ethical stance, justifiable in the consequentialist terms outlined by Clifford. Skeptics can argue that not only are they motivated by ethical considerations, but these considerations in turn are driven by a concern for human welfare. After all, if we accept Clifford's argument in its entirety, then any paranormal belief can, in principle, be a factor in propelling humans back to barbarism.

The ethical aspect of skepticism does not end there. A look through the writings of major consequentialist ethical theorists reveals that they did not simply try to make actions positive in their consequences. They tried to work out what would maximise human wellbeing. Indeed, one of the great consequentialist thinkers, Jeremy Bentham (1789), tried to derive a 'calculus of pleasure', that is a way of working out what actions would maximise human wellbeing and minimise human pain.

Thinking along these lines, it is reasonable for a skeptic to wonder whether this type of thought can be extended into skeptical ethics. That is, can the skeptic select which paranormal claims to examine, in such a way as to maximise the good consequences and minimise the bad ones? Without being as ambitious as Bentham, can we work out how to use skepticism in order to benefit humanity as much as possible?

Some simple considerations make it very important indeed. First, skepticism is completely dwarfed by the sheer scale of paranormal belief. We saw from opinion polls that a high proportion of the population believes in some form of paranormality. We also saw that a minority of believers – though still far more than the number of skeptics – holds several paranormal beliefs. Thus, there seems little prospect of skeptics persuading the vast majority of the population of their views.

The ethical argument suggests a way out of this impasse. If we accept that skepticism is to some degree an ethical position, then this ethics also suggests that as skeptics we should act so as to minimise the harm that paranormal beliefs can do, and to maximise the good that skeptical analysis can do. In the remainder of this chapter, I will sketch out what an ethics of skeptical choice might look like. I want to stress that this is in no way coercive: it provides a way for skeptics to think about their activities ethically, as opposed to simply doing what seems fun or what is most interesting.

THE BASIS OF SKEPTICAL ETHICS

The question we are addressing is: given that skepticism has an ethical component, how should we allocate our skeptical efforts so that we can maximise the good done, or avert the harm? I have considered this question in a recent paper (Bridgstock 2008), and have developed some conclusions, which I will outline here.

One way of looking at the matter is like this. There is a range of paranormal beliefs that are mostly independent of each other; a belief in ESP does not necessarily entail believing in Atlantis, nor in UFOs. Which ones should skeptics investigate, assuming that the skeptics are open to the possibility of the paranormal claims being shown to be either true or false?

As we have already seen – and as most paranormalists would agree – the chance of most paranormal beliefs being true is actually very small. It therefore follows that we should spend a good deal more effort examining the consequences of what happens if the beliefs are false. Although some paranormal beliefs may be true, most of them are surely false, and so the consequences of them being false are much more likely to occur. This means that we should look much more carefully at this eventuality.

What characteristics might lead us to believe that a particular paranormal belief – assuming that it is false – has bad consequences? In my view, there are three characteristics which make false paranormal beliefs more dangerous. One is that they are concerned with vital areas of human life: if they are believed, and are false, the consequences can be terrible. Second, it is clear that some paranormal beliefs pose a direct threat to scientific investigation. The third characteristic is that the beliefs are being 'pushed' by a powerful interest – such as a government or a major religion – and so are more powerful than most other such claims. It is worth spending a little time on each of these.

What matters are important to humans? Life and death are always important, so if we find a paranormal belief which – assuming it is false – threatens human life and wellbeing, we might decide that we wish to spend effort evaluating it and revealing its true nature. Of course, there are large numbers of paranormal beliefs which are directly connected to human life and health. The most important are the alternative and complementary medical beliefs. As we saw in the case of Liam Williams-Holloway, a decision regarding alternative medicine can be a matter of life and death, not only for people holding the beliefs, but for those who are dependent on them.

On the other hand, the Loch Ness Monster does not appear to address any such vital issues, and is less important, whether or not it exists (Sladek 1984). The same applies to other monsters such as Champ (the Lake Champlain monster), susquatch, yowie, yeti and many more.

The second criterion that might lead us to focus on particular beliefs is whether the paranormal belief directly threatens science. Science is not always right, but it is an extremely valuable human method of working out how the universe works, and its benefits to humanity have been immeasurable.

We have already seen that some of the founders of modern skepticism were disturbed by the threat to science that some paranormal beliefs pose. Creation science, as we have seen, actively seeks to alter the very notion of what science is – it should, in the view of the creation scientists, be subordinate to particular religious beliefs. Because science is so important, both in influencing our view of the universe and in generating new technologies, it is clear that we can regard creation science as a belief that should receive close skeptical scrutiny.

Finally, there is one other danger signal which suggests that skeptics might become involved as a matter of urgency. This is where some powerful group or organisation appears to be supporting paranormal beliefs. The reasoning here is fairly straightforward: if a belief is being promoted by a powerful group, then it has a much better chance of acceptance, and so, because of the consequences of this, it would be better if it were properly evaluated at an early stage.

A few examples may clarify the matter. The Melbourne Metropolitan Water Board, a statutory authority in the state of Victoria, hired dowsers to help it find new supplies of water (Plummer 2004b [1986]). Although the validity of dowsing is not by itself a matter of life and death, the Melbourne Metropolitan Water Board was using taxpayers' money to hire the dowsers, and so more caution would be expected when using unproven methods. In particular, if statutory authorities start to use paranormalists as part of their normal functions, this gives legitimacy to these practitioners which they can use to garner more respect elsewhere. It is certainly true that many paranormalists, such as psychics, are very anxious to achieve respectability through these means, and many psychics claim to have assisted the police in finding the bodies of murdered people.

There are good reasons for paying special attention to paranormal practitioners or beliefs endorsed by important or governmental bodies – with or without the payment of money. Being privileged in this way gives the practitioners a cachet, and a perceived validity, which they might not otherwise enjoy. Thus it would be perfectly legitimate for skeptics to become interested in beliefs that are privileged in this way.

TWO TERRIBLE EXAMPLES

We might look at two examples where all three of the criteria are fulfilled. Both have gone down in the annals of infamy. They are the use of 'Aryan Science' to justify the persecution of the Jews in Hitler's Germany, and the appalling consequences of Lysenkoism in the Soviet Union.

'Aryan Science' in Hitler's Germany had all the characteristics of a pseudoscience. The origins of the scientist mattered more than the quality of the science, so relativity was dismissed as 'Jewish Science' (after all, Einstein was Jewish, so his work must be suspect!). Scientists – even those in universities – were closely supervised, and those of Jewish descent were driven from their positions. The result was a catastrophic decline in the quality of German science. David Hilbert, the veteran mathematician was asked after a few years of these policies, about the state of German mathematics. His reply was: 'It just doesn't exist any more!' (Jungk 1965, p. 44)

More horrendous than the imposition of 'Aryan Science' was the use of this pseudoscience to justify the persecution of 'inferior' races. It was argued that certain groups, such as Poles, Gypsies and above all Jews, were genetically inferior, and so either had no right to reproduce, or no right to life. This was part of the justification for the concentration camps and, ultimately, the murder of millions of people by shooting, gassing, starvation and disease. It is fair to describe this episode as one of the darkest and most shameful stories in human history – and it was justified by pseudoscientific beliefs (Shirer 1964, p. 1165).

Of course, the pseudoscience was only part of the picture. The pathological anti-Semitism of Hitler and the other Nazis was not caused by the pseudoscience: the pseudoscience was used as a justification. By our criteria, of course, this should sound a triple warning. The issues involved were those of life and death (they involved the right of some groups to exist), they threatened the operations of science, and they were being pushed by a powerful entity, the German government. The results were the entire disgusting edifice of Nazi science. Millions of people lost their lives in the concentration camps, and tens of thousands were tortured in the pseudoscientific 'experiments' of Dr Mengele and his colleagues. Left to themselves, the Aryan Science proponents would have been an unpleasant set of cranks. With the backing of a totalitarian government, the impact of their ideas became quite monstrous.

Our second example is less dramatic, but equally bad in its effects. This was the rise of Lysenkoism in the Soviet Union and, later on, in communist China. Trofim D Lysenko was an agricultural scientist who attracted the support of the Soviet dictator Stalin. He made sure that Lysenko gained great power and prominence, and that all who opposed him were neutralised. In particular the agricultural scientist Nikolai Vavilov lost his life as a consequence of his outspoken opposition to Lysenko (Medvedev 1978).

By itself, this is bad enough. However, the real problem was that Lysenkoism was a disaster for Soviet agriculture. It involved a series of strange propositions, such as the idea that genes do not exist, and its ideas when translated into practice meant that Soviet agriculture was damaged for decades. Since one of the characteristics of the Soviet Union was its inability to provide enough food for its people, it is clear that this pseudoscience, backed by Stalin's power, caused great hardship. It concerned vital issues, it attacked science, and it was supported by a powerful dictator (Medvedev 1978).

The story is worse than this. Lysenko's ideas were also taken up by the Chinese communist dictator, Mao Tse-tung, and imposed upon hundreds of millions of people during the so-called 'Great Leap Forward'. The damage to the food production of China was immense, and it seems likely that several million people died as a result (Becker 1996).

Although there is a good deal more to be said on this subject, the basic lines of thought should be clear enough. If we accept that we have any kind of ethical duty to fit our beliefs to the evidence, then it also seems to follow that we have an equivalent ethical duty to ensure that our skeptical efforts are focused on those paranormal beliefs where the gains are likely to be greatest. As a first approximation, it seems likely that three criteria should be used to sort out which paranormal claims should receive most attention. One criterion is that paranormal claims should involve areas that are important to humanity, and the extent to which the truth or falsity of the beliefs have implications for human welfare. Another consideration is whether the operations of science are under threat. The final criterion is that special concern should be aroused if there are signs that the paranormal beliefs are being actively supported by an important institution or agency. The reason, of course, is that such privileged beliefs are likely to be far more important for humanity than others, and so their truth or falsehood should be established as soon as possible.

Other examples of these criteria can be applied to other cases, although not as terrible as the ones just mentioned. One such example is the case of the Queensland government, in the 1980s, which actively began to support the teaching of creation science in state schools (Knight 1986). This, under the outlined criteria, would qualify creation science as a high priority for skeptical investigation.

It seems that our extension of ethics into skeptical activity has had some unexpected consequences. If we accept that skepticism has an ethical component, then it seems logical that this component should affect the paranormal claims – and other claims too, for that matter – which we investigate. Assuming that the claims are false, which they usually are, then we have found some useful guidelines which can assist us with deciding what claims to investigate and what claims are not worth looking into. Of course, the criteria could be extended. If a paranormal area is already receiving a good deal of skeptical attention, then we might legitimately decide not to become involved, but to concentrate on another area, less urgent but less well covered by skeptics. Of course, another consideration altogether is whether our particular expertise is more suited for one area of study rather than another. The consequentialist approach is not hard and fast, but it can assist us with making decisions about what to investigate and what not to worry about.

We have seen earlier that CSICOP has changed its name to CSI, the Committee for Skeptical Investigation, and has broadened its remit to include the investigation of controversial or extraordinary claims generally (Frazier et al 2007; Kurtz 2007). This was criticised in a passionate article by Daniel Loxton (2008) on grounds similar to those advocated here. Loxton argues that someone has to be concerned with the proliferation of paranormal claims, and that only CSICOP is. Therefore, he believes, skeptics should focus on the paranormal – in particular prevention, harm reduction and trying to bring justice – because no one else is primarily concerned with this area.

The force of Loxton's argument is clear. If we neglect the investigation of the paranormal, or pay it less than due attention, then dreadful consequences of the kind outlined out in the Introduction will become more common. At the same time, there may be non-paranormal claims which can usefully be analysed, and where the effort can be justified ethically. Perhaps some sort of a compromise is possible, where the main focus of skepticism remains on the paranormal, but some investigation of non-paranormal claims can be made where skeptical expertise is appropriate. We will take up the question of how far skeptical investigation can be applied to the non-paranormal in Chapter seven.

Ethics also offers us useful guidance in other areas of skeptical activity. Not only does it enable us to decide what to study, it also enables us to examine some of the activities we might indulge in, from a consequentialist ethical viewpoint. Consider the topic of hoaxing, for example, which is often regarded as a lethal weapon used by skeptics against credulous paranormalists.

To show the kinds of issues that can arise, we will look at one aspect of skeptical conduct: when is it ethically appropriate for skeptics to hoax or fake results?

SKEPTICS' USE OF HOAXING AND FAKING

Occasionally, skeptics may face the possibility that they can test the effectiveness of a paranormal claim by deceiving the claimants. That is, they may be able to perpetrate a hoax. With a hoax, the aim is to expose the deception at some time, with consequent loss of credibility to the claimant.

Thomas F Monteleone perpetrated a simple example of hoaxing, and he described it in an article in *Omni* (Monteleone 1979). A radio talk show had as a guest a man called Woodrow Derenberger who claimed to have visited the planet 'Lanulos' in the 'galaxy of Ganymede'. (Of course, astronomers know of no planet called Lanulos, and Ganymede is a moon of Jupiter, not a galaxy.) Monteleone, then a student, phoned in to the program and announced that he too had been to Lanulos. However, he contradicted Derenberger's account on each point. What happened was rather unexpected:

> on each occasion he would give ground, make up a hasty explanation, and in the end corroborate my own falsifications. He even claimed to know personally the 'UFOnaut who contacted me!'
>
> (Monteleone 1979, p. 146)

This is an amusing student prank. However, Monteleone found that it would not go away. At the time of writing, the consequences of his deception had haunted him for twelve years. People kept contacting him about his UFO experience, and he was introduced to the 'odd, achingly pathetic world of the UFO cultists'.

> They usually phoned very late at night and spoke in nervous whispers, claiming to be fearful of the FBI and the CIA, who were always trying to bug their phones... When they came to my apartment to interview me, they always traveled alone, lived out of beat-up suitcases, and seemed to have an affinity for Radio Shack portable tape recorders.
>
> (Monteleone 1979, p. 146)

Monteleone told these 'investigators' even more absurd stories, wanting to see how far he could push their credibility. However, this did not work:

> To my horror, I soon realised that I could have told these 'investigators' that the aliens had taken the Washington Monument to Lanulos as a souvenir, having replaced it with a fake back in 1956... and they would have believed me!
>
> (Monteleone 1979, p. 146)

The people Monteleone met certainly seem to have been a sad bunch, desperately looking for proof that UFOs are visiting us, and largely incapable of assessing the evidence. Monteleone's conclusion points to the 'deep psychological need that many of these people have – a need to believe in something greater than themselves'. Monteleone (1979, p. 146) This, he thinks, is what makes them so pitifully gullible.

This is a good story, both sad and funny at the same time. Quite inadvertently, Monteleone showed the poor research standards of at least some ufologists, and also threw light on some of their deeper motivations. However, Monteleone's story takes us into the ethical realm. When exactly is it ethical to perpetrate hoaxes like this, and when does it go beyond proper behaviour for a skeptic? After all, although Monteleone did not plan to do so, he actually caused some negative consequences. It sounds as if many of the UFO people were short of money, yet they travelled long distances to talk about Monteleone's 'UFO experience', and presumably incurred considerable costs as a result of this. Yet the subject of their investigations was a fake. Again, Monteleone's 'experiences' on Lanulos may have strengthened the saucer movement: apparently UFO enthusiasts passed his name around as someone with important evidence about the aliens. Should he have done this? Was it right or wrong?

A simple and robust answer would go along these lines. UFO claimants, like all paranormalists, hold beliefs on the basis of insufficient evidence. Thus, according to Clifford, they are breaching some basic ethical principles. Furthermore, they are publicly propagating their claims and stating that they are true. Therefore, hoaxing is simply an effective method of demonstrating that they do not know what they are talking about: it is both effective and ethical.

Still, UFO people are human beings like the rest of us. Having strange ideas does not deprive them of their basic humanity. Monteleone knew full well that he was giving them false information, yet he continued to do so. And, in the process, he probably increased the number of people believing on the basis of inadequate evidence.

On consequentialist grounds, we might consider what the costs and benefits of Monteleone's actions were. On the one hand, he discovered how gullible some UFO researchers are, and passed this knowledge on to other people. On the other hand, he appears to have given the movement something of a boost by apparently substantiating Woodrow Derenberger's accounts, and adding to the list of 'UFO contacts' circulating among the ufology fraternity. On balance, although it is not completely clear, if I had been Monteleone considering the prank, I would probably not have done it. Alternatively, I might have performed the original hoax, and then rapidly revealed it to be bogus, thus nipping the use of my example in the bud. The major consequence of the action appears to have been a strengthening of an already irrational movement.

We should perhaps notice one disconcerting feature of this little story. The evidential standards of the ufologists – at least according to Monteleone – are so poor that even a hoax becomes press-ganged into the role of evidence for their beliefs. From a consequentialist viewpoint, this is a major negative feature when considering the ethics of hoaxing. Anyone contemplating a hoax involving paranormal claimants must remember that a major consequence may be a strengthening of the evidence for the belief.

HOAXING THE PARAPSYCHOLOGISTS

Let us consider another case that attracted far more publicity and acrimony. It was far more deliberate in nature, and organised by a leading skeptic. In 1979, Washington University, St Louis, announced that it had received half a million dollars to work on paranormal research over a period of five years. A physicist, Peter R Phillips, was the chief researcher, and the work of the lab focused on whether psychokinetics (notably spoon-bending) could be carried out under laboratory conditions (Randi 1983a).

Over several years, researchers began to be excited by the performance of two young men, from different states, who appeared to show some of the metal-bending abilities of Uri Geller. These two young men, Steve Shaw and Michael Edwards, were apparently able to perform tasks such as blowing fuses, turning a rotor under a glass dome and moving a small alarm clock. Perhaps the most astonishing instance was when metal objects were bent in the lab overnight, when nobody was supposed to be there.

At the same time as these amazing events, James Randi wrote repeatedly to Phillips expressing concern about the ability of the researchers to guard against fraud and offering to help improve the project's safeguards. These offers were not taken up, and in 1981 the researchers presented a paper to the Parapsychological Association. Other parapsychologists were critical of the results and the researchers decided to withdraw the paper.

About this time rumours began to circulate that Edwards and Shaw were conjurers, and early in 1983 Randi held a press conference at which he announced that Edwards and Shaw were in fact conjurers whom he had planted to show the defectiveness of the safeguards of the project.

In later papers, Randi outlined his view of the project. He termed the deception 'Project Alpha', and spelled out his view of the essential flaw in most paranormal research, and what he expected to happen:

Parapsychologists have been lamenting for decades that they are unable to conduct proper research due to the lack of adequate funding, but I felt strongly that the problem lay in their strong pro-psychic bias. The first hypothesis, therefore, was that no amount of financial support would remove that impediment to improvement in the quality of their work . . . the second hypothesis was that parapsychologists would resist accepting expert conjuring assistance in designing proper control procedures and, as a result, would fail to detect various kinds of simple magic tricks.

(Randi 1983a, pp. 24–5)

Both of Randi's hypotheses have received support from other researchers. For example Susan Blackmore, a parapsychologist turned skeptic, has written of the strong commitments to paranormal belief that most parapsychologists appeared to have (Blackmore 1996). And, judging by the results of Project Alpha, both of Randi's hypotheses were strongly confirmed. The experiments in the St Louis lab were not well enough controlled to stop the young magicians from wreaking havoc, and the offers of help from Randi were not taken up in time to prevent the fraud from taking place (Randi 1983b).

Michael Thalbourne, who was involved with some of the St Louis research, has argued that Randi's interventions were unethical, and that the project was still at an exploratory stage. He distinguishes between exploratory and formal research, stating that this distinction is common in parapsychology:

... formal conditions are generally more difficult, more time-consuming and usually more costly to assemble than are informal conditions. One would be misspending one's resources to insist upon setting up those formal conditions – which may for example involve expensive equipment – before getting some preliminary idea as to whether the information one is after gives some appearance of being present.

(Thalbourne 1995, pp. 351)

This is reasonable up to a point. In many fields of research, preliminary probes are useful and necessary. Survey researchers may send out a few 'pilot' questionnaires to see if the questions make sense. Scientists with complex research devices will often do preliminary runs of their experiments, to satisfy themselves that the equipment is working properly. In the same way, it would be perfectly acceptable for parapsychologists to do a little exploratory work with subjects who seem to show some promise.

However, no scientist would present the results of preliminary trials as though they were the final result, and no social scientist would present the results of a pilot as though they were final. At the very most, researchers might explain their methods of preparing for the research. The St Louis researchers did present a paper on their results, and apparently were quite surprised when other parapsychologists were critical.

Furthermore – and this is a key point – parapsychology has a sad history of fraud. In this field, above almost any other, it would seem foolhardy for any researcher to present results where fraud could possibly be present. Yet this is what these researchers did, and they paid a big penalty.

What about the publicising of the hoax? Randi let the magicians spend several years worming their way into the organisation, and went public as soon as they had made their point. Thalbourne's view is that, after perpetrating the hoax, Randi should have approached Phillips and the other researchers, and written a joint paper on the event (Thalbourne 1995). This assumes that Randi and the parapsychologists are members of the same research community. However, Randi's assumptions are precisely that the parapsychologists were not proper researchers, and also that they were the subject of his research, not co-researchers. These were the assumptions from which his hypotheses flowed. Therefore, it is not at all clear that Randi was ethically bound to share his results with the parapsychologists.

My own assessment is that Randi's hoax was within ethical boundaries. Consider it from a consequentialist perspective. It was devoted to showing an important fact about a well-endowed parapsychology project, it was announced as soon as the point had been made, and it does not seem to have propagated the misunderstanding it set out to correct. Randi's approach showed that even highly prestigious – and well-funded – parapsychology is prone to hoaxing, and that only much tighter controls can guard against this. He also avoided the dire pitfall of actually propagating the belief he was out to expose. There was certainly some unpleasantness, but Randi had offered explicit warnings and prescriptions about what could be done to prevent his hoax.

OTHER HOAXES

There is a long history of hoaxing in paranormal matters, some carried out by skeptics and some not. One of the funniest was perpetrated in *Cosmic Voice*, mouthpiece of the Aetherius Society. The leader of this Society, George King, claimed to be in touch with assorted important entities, and on occasion he and his followers claimed to have saved the Earth from terrible destruction.

King and his followers were apparently most excited when reports appeared of scientists who were interested in the work of the Society. These developments were reported in *Cosmic Voice*. Eventually, a rival publication, *Psychic News*, pointed out that the scientists had quite absurd names, such as Egon Spunrass, RT Fischall and N Ormuss (Sladek 1978). They finished their exposé with a question, formed from the name of another scientist: 'Huizenass?' Apparently George King was extremely angry at the hoax (Moore 1972, p. 116).

This hoax was plainly successful at exposing the pathetic lack of any critical judgement of the members of Aetherius Society – or at least of the editors of *Cosmic Voice*. The Aetherius Society is still in existence, and does not appear to have either benefited or suffered from the hoax. We really know too little about this hoax to pronounce on its ethicality. Would the author have eventually revealed what was happening – or did this happen anyway, via a tip to *Psychic News*? From a consequentialist viewpoint, we cannot find much evidence on which to make a decision.

Finally, we should look at the amazing phenomenon of crop circles and its associated science of cereology. The origins of cereology are obscure, but the matter first came to public attention in about 1980. Over the next ten years literally hundreds of circles – often part of far more complex patterns – appeared in crop fields. Although the south of England was the main site, circles also appeared in Canada and Japan.

The science of cereology developed apace. Books and journals appeared. There were several different schools of thought about exactly what caused the pattern. Jim Schnabel (1993) has written a detailed account of the hysteria over crop circles. Assorted societies with pretentious names were set up – the United Bureau of Investigation, the CCCR and the Crop Phenomena Researchers – and several books were published, including one which entered the British bookselling top ten.

Repeatedly, cereologists declared that the genuine circles could not be made by humans. The crops were not broken but bent, and there were effects such as layering and plaiting, which enabled the experts to distinguish attempts at faking.

Finally, the story was destroyed in 1991. Doug Bower and Dave Chorley (immediately called Doug and Dave) revealed that they had been creating crop circles and other patterns since about 1976. They showed the various tools of their trade, and established that many of the circles – not all – had been created by them. Doug and Dave had lingered around the circles after their creation, taking enjoyment from the strange posturing of the cereologists and mystics.

Then others came forward. The Wessex Skeptics (Hempstead 1992) had also fooled the cereologists with a few circles. Finally, late in his book, Schnabel confesses that he too has been a crop circle faker. He used a garden roller, a plank and rope and his feet to create the phenomena which the cereologists so glibly proclaim to be beyond human ingenuity.

> ... I am unable to repress a few frank memories of circles swirled by myself or by my artistic colleagues: for instance, that set of big circles at Lockeridge, formed in lonely, lofty moonlit green barley with a garden roller and some string, and energies which were still powerful enough, a day later to cure an old woman of her arthritis pains ... The triplet near the town of Wroughton, so well-wrought that George Wingfield said of it 'If this is a hoax, I'll eat my shirt in public.' ... The rough-hewn ring beneath which, according to the newspapers, an expert dowser detected a buried stone circle.
>
> (Schnabel 1993, pp. 284–5)

The effect of these revelations was a collapse of the whole field of cereology. They moved to other occult topics, or back into mundane reality.

What are the ethics of what was going on? As far as it is possible to tell, the Wessex Skeptics confined themselves to a few circles, and did not hesitate to reveal that these were faked as soon as possible. On the other hand, Doug and Dave manufactured circles year after year, taking great amusement from the antics of the cereologists and made no attempt, until right at the end, to show how ill-equipped the cereologists were to make sense of the phenomenon. In a very real sense, it seems that without the hoaxing, the field of cereology and all its paranormal accoutrements might not exist.

We cannot judge Doug and Dave by the same standards as we might judge skeptics. Apparently they did hold some paranormal beliefs, and were simply two private citizens with a rather unusual – and amusing – hobby. They actually propagated the myths which they despised so much, and held their silence year after year as the movement grew. It is likely that without their intervention there would have been no crop circle craze, or at least one of much smaller proportions. Since skeptics are devoted to investigating paranormal phenomena, and presumably therefore opposed to fake paranormal phenomena, this would be unethical for them. The consequences of the hoax were quite dire, though rather funny.

There seems to be no clear message from these hoaxes. Some of them clearly contributed to belief in paranormal phenomena, others made little difference. The Randi and cereology hoaxes did eventually cause the collapse of some paranormal beliefs, though it is not clear whether the latter would have existed at all without Doug and Dave's faking.

Perhaps these simple ideas best summarise the position. First, any hoax is likely to produce both positive and negative results. The paranormalists may be shown to be incompetent in their efforts to sort out what is true and what is not. It may also be possible to show that the paranormal phenomena are themselves the result of fraud in some way. On the other hand, the hoax may well produce a good deal of personal grief for the people who are hoaxed, and it may blight their lives. Further, and most surprisingly, a hoax may actually fuel the phenomenon which it is intended to undermine.

Clearly, in order to work out whether a hoax is ethical or not, we should be satisfied that the good effects will outweigh the negative. The hoaxes of Monteleone and Doug and Dave actually seem to have fuelled the UFO and crop circle crazes. Indeed, in Doug and Dave's case, the hoaxers actually seem to have created the craze. If the aim is to stop the craze by showing it to be false, the track record is not impressive. The only paranormal field which has been demolished by the revelation of the hoax was cereology. The Aetherius Society, ufology and parapsychology do not seem to have been seriously affected by the hoaxes. Therefore, it seems unlikely that a revelation of incompetence is going to have any effect on a paranormal belief.

It follows that as a first requirement paranormal hoaxers should not proceed where they believe that there is a chance that their dishonesty will propagate the fad rather than halt it. Secondly, the mere revelation of incompetence will not halt the fad. These are important considerations, which might give prospective hoaxers serious pause for thought.

ETHICS AND PARANORMALISTS

I attended my first ever Australian Skeptics convention in 1986. One speaker detailed some fairly blatant faking by clairvoyants. This angered a clairvoyant in the audience, who stood up and declared his identity, before continuing, 'But if I am to be true to myself, why should I take tests organised by the skeptics?'

There is an answer to the clairvoyant's question, and it is ethical in nature. As we have seen, almost any paranormal proposition, if true, has the potential to revolutionise humanity's position. Clairvoyance, for example, if shown to be true, will demonstrate once and for all that humans are not simply material creatures, but have a spiritual dimension: there will be no more room for doubt. In addition, as we have seen, the implications of applying clairvoyance are quite staggering.

The answer to the clairvoyant's question is, therefore, that anyone who seriously believes that they can demonstrate paranormal powers can revolutionise human life. Therefore, there is an ethical obligation to take any necessary tests to convince those who doubt. The reward for humanity is so great that any paranormalist who shirks the task could be regarded as behaving unethically.

I have tried this argument out on a few paranormal believers. To me it is devastating: it suggests that people who claim paranormal powers but do not accept testing should be regarded with a certain amount of contempt. However, I have found that paranormal believers do not seem to be able to grasp this point. I do not know about the uncommitted, but it does seem a reasonable weapon for skeptics to use in public debate: if the paranormalists are so sure of their powers, why not accept rigorous testing? That way they can prove their case and benefit humanity. They will probably make a fortune for themselves in the process!

ETHICS AND SKEPTICISM

It should be clear by now that important ethical dimensions exist for skepticism and the paranormal, and that human lives and welfare can be affected by what is done. In our examples of the ethics of belief, the selection of claims for ethical attention and the hoaxing of paranormalists, an ethical perspective causes us to see these matters in a rather different light. In my view, too, the ethical argument for paranormalists to accept testing is extremely powerful.

It seems clear that although Clifford may have exaggerated his arguments for an ethics of belief, holding ill-supported convictions can result in disasters, and can also corrupt our ability to assess warranted from unwarranted beliefs. The concern for disasters should also have some effect on our selection of skeptical areas for study. Some paranormal beliefs are inconsequential, others can result in dreadful suffering. Finally, when we deal with the possibility of hoaxing paranormalists, we should be clear in our own minds that what we are about to do will have beneficial effects.

LAUNCHING PAD

Because the area of ethics and skepticism is so under-researched, there is almost no material that is directly related to the topic. Clifford's (1879) paper is clearly written and is a good place to start. Two recent papers by Zamulinski (2002) and Degenhardt (1998) comment on Clifford's ideas and some of the controversy.

More generally on ethics, it is useful to see how philosophers actually think about these matters. Peter Singer has written many important books, but one (Singer 2001) summarises his thought in a range of areas. The fact that his approach is broadly compatible with that of Clifford is a bonus, though prepare to be disconcerted as Singer thinks through the implications of his assumptions. More modestly, the papers by David Koepsell (2006) and myself (Bridgstock 2008) make a start in the field of ethical skepticism.

7 | Skepticism beyond the paranormal

W E HAVE SEEN that skepticism is a powerful tool in evaluating paranormal claims. We have started with a set of important skeptical principles, and have argued that these are crucial in grasping the nature of the argument. Between them, the burden of proof, Occam's razor and Sagan's balance provide the kernel of a skeptical approach to all aspects of the paranormal.

In addition, we have seen that a range of intermediate concepts are needed to link the general principles with the paranormal claim. For example, claims that an alternative therapy can cure diseases – with an amazing story of how someone was cured – can lead to discussion in terms of the placebo effect, and the need for double-blind controlled experiments to see whether the treatment actually works. The burden of proof is clearly on the claimant. Since the claim is an amazing one, Sagan's balance suggests that the quality of evidence must be extremely high. Finally, if the placebo effect can explain the results, there is no need for the paranormal explanation at all. In this way, thinking about the paranormal can be revolutionised by the use of these basic principles, and the intermediate concepts.

Paranormal ideas are only a small fraction of all ideas in the world. In this chapter we ask an important question: have skeptical approaches any value outside the field of the paranormal?

Psychologically, the answer to this is very clear. Every year, after my course, students tell me that they are more skeptical about newspaper reports, politicians' statements and commercial advertising as a result of taking the course. One student, for instance, wrote to me and said:

I wanted to thank you for teaching my brain to question and rationalise, as opposed to blind and unquestioning acceptance; it will help me greatly in my BA in Politics! There has been many an occasion when I have seen a program on television and have wanted to call you up and discuss it! I watched a program on the ABC's *Catalyst* a few months back about frontal lobe epilepsy and how it can trigger religious experiences with great interest. And what's with the new wave of television criminal psychics?

Clearly, this student is greatly enthused about her ability to think critically about the paranormal. In addition, she sees that her new skills may be useful in non-paranormal areas, such as a BA in politics. Still, just because people *do* something, it is not necessarily justified. Perhaps there are good reasons for not applying skepticism to areas outside the paranormal. And, indeed, there is a strong logical barrier to using skepticism for other than paranormal claims.

The barrier is simply this. By its definition, as we saw in Chapter three, skepticism consists of applying scientific and related perspectives to claims of the paranormal. Therefore it is logically impossible for it to be used on non-paranormal phenomena. Once we stop thinking about the paranormal, we have stopped using skepticism.

We have to concede the force of the logical argument. Strictly, even the most meticulous use of skeptical criteria on non-paranormal beliefs cannot be skepticism. The student above clearly saw a connection between skepticism and other areas of thought. So do the Australian Skeptics. Their magazine has articles arguing about the validity of the greenhouse effect (Lowe 2002), the value of nuclear power (Keay 2001a; 2001b), pyramid selling schemes (Lead 1998), linguistic theories (Curtain and Newbrook 1998) and aboriginal cannibalism (Buchhorn 1999). The American skeptical magazine *Skeptical Inquirer* is less varied, but does print articles on non-paranormal items such as graphology (Tripician 2000), children's behaviour (Fiorello 2001) and why Darwin was buried in Westminster Abbey (Weyand 2006).

As we have seen, the major American skeptical organisation has now changed its name to CSI, and wishes to extend its investigations beyond the paranormal (Frazier et al 2007; Kurtz 2007). This has been criticised by Loxton (2007). His view is that we continue to be vastly outnumbered by paranormal proponents, and that somebody has to take responsibility for investigating and debunking these claims. We have looked briefly at the ethical issues involved. This chapter deals with the question of how far it is feasible to use skeptical approaches beyond the boundaries of the paranormal.

USING SKEPTICISM OUTSIDE
THE PARANORMAL

How might we go about using skeptical approaches on non-paranormal beliefs? We have seen that the paranormal, by its very definition, consists of claims that lie outside normal knowledge and scientific belief. By analogy to science, there are bodies of knowledge and belief which are also generally accepted, but which are not scientific. There are academic bodies of knowledge, such as those in history, literature and economics.

These bodies of accepted knowledge are backed by differing levels of evidence. Like science, none are certain. Some, like science, are carefully evaluated and constantly reviewed, so that their strongest findings would take a great deal of evidence to refute. Others are based on little more than myths or anecdotes.

Clearly, wherever there is a body of generally accepted knowledge or belief, there can be dissent from it. This dissent is analogous to paranormal belief, in that it questions existing knowledge. If we are to proceed in a skeptical fashion, our attention should be focused on the claims of those who dissent from the established body of knowledge. Can the three principles be applied to these different forms of dissenting thought?

For paranormal claimants, the three principles present a daunting obstacle. Since paranormal claims are, by definition, scientifically implausible, it is often hard for supporters to make a strong case. However, many accepted beliefs are less strongly supported than scientific theories, and indeed some have almost no supporting evidence at all. Thus, instead of having to overturn a mass of scientifically established evidence, the claimant may be faced only with the task of producing some sort of support.

Consider, for example, the famous *MythBusters* television show, in which Adam Savage and Jamie Hyneman set out to test an assortment of myths. In one show (*MythBusters* 2003) they tested a whole collection of myths to do with the breathalyser machines used by the police to catch drunken drivers. Among other myths, some people believe that holding a coin or a battery in the mouth will fool the breathalyser. Others hold that eating onions will do the trick, and yet others that coughing will clear alcohol fumes from the lungs. Adam and Jamie drank enough alcohol to put themselves over the limit, and then tried all the avoidance tricks, aided by a policeman and a breathalyser machine. Nothing worked.

What should we make of this? Breathalyser machines are widely used in many countries, so presumably there is reason to regard them as effective.

On the other hand, we also know that government and other authorities can make wrong decisions. Indeed, sometimes politicians make decisions simply because they want to be seen to be taking action, even if the action is ineffective. This is satirised in shows such as the British television series *Yes Minister* (Lynn and Jay 1983). What is more, although a breathalyser is based on scientific principles, a way of beating it might not contravene those principles: it might simply stop the machine working effectively.

A reasonable conclusion might suggest that the skeptical principles can be applied to cases like the breathalyser, but with reduced force. We might lean to accepting the validity of breath test results, unless believable evidence shows that the machines do not work. Following Sagan's balance, we would require less evidence to reverse our belief than for paranormal claims. Maybe Adam and Jamie's experiments, if they had worked, would have provided enough.

Outside the area of the paranormal, Occam's razor certainly applies: we should generally prefer the simpler explanation, one that involves the fewest entities, and preferably no new one. However, unlike the case of the paranormal, this is sometimes quite possible, and history is full of examples where it has happened.

Consider, for example, Copernicus. He was one of the first Europeans to argue that the Earth went round the Sun. Until that time, the predominant belief had been that the Earth stood at the centre of the universe, and all the stars and planets revolved around it. The Church had adopted this idea as part of its doctrines, and so people dissenting could run into profound trouble. Very prudently Copernicus arranged that his book – which argued that the Earth went around the Sun – should be published on the day of his death. As a result, he escaped any persecution, but his ideas were propagated through the community of astronomers.

Progress was slow, but Copernicus's ideas had a huge advantage over the ideas of Ptolemy, the great Greek astronomer. Whereas Ptolemy's system required three hundred epicycles – circles mounted on circles – to explain the movements of the stars and planets, Copernicus's system required only about thirty: it was much more simple. It took more than two centuries for the ideas of the Church and the ancient Greeks to be fully overthrown, but eventually it did happen (Kuhn 1957). The simplicity of the newer ideas gave them a great advantage.

Let us look at a current non-paranormal example, where economy of explanation is the crucial deciding factor. This is the notorious case of the *Apollo* Moon 'hoax'.

THE GREAT APOLLO MOON 'FAKE'

You may have seen the movie or read the book *Capricorn One* (Ross 1978). It is a good story. The people who have been charged with putting Americans on the Moon are in a desperate race with the Russians. However, they have realised that they are unable to succeed. So the astronauts are taken to a secret location, and there act before cameras on film sets made to look like the Moon. In short, the American Moon landing is a fake.

It is a possible scenario. Of course, there is a nasty implication. The astronauts are not needed after the hoax. Indeed, they are a positive menace to the success of the enterprise, and have to be eliminated.

For some people, the *Capricorn One* scenario is not fiction. In their view, the *Apollo* Moon missions were also a fake, with the famous pictures of Armstrong and Aldrin on the Moon being shot in a studio here on Earth (Plait 2002).

We can see at once that this is an enormous conspiracy theory. Many thousands of people were involved in the *Apollo* project, and all of them must have taken part in a huge conspiracy to mislead the public. What is more, it is an international affair, as the Australians who received the first transmissions from the Moon must also have been complicit in the deception.

Now, there is an immense body of knowledge supporting the view that American astronauts went to the Moon. The event was extensively televised, and written about in books and newspaper reports. Rocks and photographs were brought back. The astronauts themselves talked about their experiences and feelings. What is more, tracking stations throughout the world, including Australia, picked up the *Apollo* transmissions.

In addition, other conspiracies, such as the Watergate cover-up, the Mafia and the plot to exterminate the Jews, have yielded people who eventually repented and described in detail what had happened. Nothing like that has occurred in this conspiracy. In addition, as we shall see, the conspiracy was amazingly stupid. The evidence is blatantly false, or so the theory goes.

As we have stressed throughout the book, the theory stands or falls on the evidence. What evidence is there that Neil Armstrong and Buzz Aldrin never went to the Moon? Almost all of the evidence comes from an examination of the pictures which are claimed to have come from the Moon. It is claimed that they have properties which show clearly that they are fakes. Rather than go through all the evidence, we will simply look at a few key points. Skeptical astronomer Philip Plait (2002) gives an exhaustive analysis.

A widely touted piece of evidence often produced is that when you examine photos from the *Apollo* missions, no stars are visible in the sky. Does that not show that the astronauts were not on the Moon? Well no, not really. Cameras nowadays focus on the brightest part of the scenery, and the light aperture adjusts so that this is photographed. Clearly, the brightest part of the picture is likely to be the astronauts, their craft and equipment, and so the camera admits this light. In doing so, it necessarily excludes the much dimmer light of the stars (Plait 2002, pp. 158–60).

Another claim of the *Apollo* hoaxers is that the shadows in the photographs are not parallel. Since there is only one source of light on the Moon – the Sun – there is clearly something wrong if the shadows are non-parallel. Phil Plait pours a good deal of scorn on this line of argument. After all, if NASA was setting out to fake lunar pictures, would it not make sure they look authentic? In fact the impression that shadows are non-parallel can also be seen on Earth. Plait points out that railway lines do not appear to be parallel when we look along them, even though they must be. He continues:

> The same thing is happening in the lunar photographs. The shadows don't appear to be parallel because of perspective. When comparing the distance of shadows from two objects at very different distances, perspective effects can be quite large. I have seen this myself, by standing near a tall street lamp around sunset and comparing its shadow to that of one across the street. The two shadows appear to point in different directions. It's actually a pretty weird thing to see.
>
> (Plait 2002, pp. 171–2)

After reading this, I went out at sunset and had a look at shadows from two street lamps near my home. Plait is right: the effect occurs here on Earth, and is easy to see if you bother to look for it.

For an amazing claim like this, the burden of proof clearly lies on the claimants. It is accepted as a historical fact that people went to the Moon, so we must look for evidence to convince us otherwise. As we have seen, there is a mass of evidence supporting the reality of the *Apollo* missions. The claim that they are fakes is not paranormal, but we would want some pretty strong evidence to tip Sagan's balance and convince us of a huge conspiracy.

At this point Occam's razor becomes crucial. The conspiracy claimants have produced some points from the photos which, on the face of it, look convincing. Plait, however, has succeeded in explaining all of the claims of the hoax proponents by using known processes, and so by Occam's razor he has eliminated the need for a conspiracy explanation. After all, by explaining the features of the photos in terms of factors like perspective, features of cameras and the Moon's environment, Plait has used existing concepts to explain what we see: we don't need a huge new concept like a monstrous, watertight but extremely stupid conspiracy.

In sum, it looks very much as if Occam's razor and the other principles apply just as readily in the non-paranormal area as they do to the paranormal. There is a difference, however. The difference is that paranormal claimants face the problem that their explanations are almost always less economical than any natural explanation. By contrast, it is perfectly possible that a dissenting explanation in the non-paranormal area can be more economical than the established explanation. Copernicus managed this, and so can other dissenters, though they still face the natural human tendency to stick to what we already know.

We have seen that the three skeptical principles work quite well in the non-paranormal area, though sometimes they have to be used somewhat differently. This is usually because the evidence against dissenting claims outside the paranormal area is less overwhelming than within it, and so generally there is less the onus on the dissenter. As an example of how this plays out in practice, let us look at a notorious case of dissent in a non-paranormal area, where feelings have run high and the logic of evidence has swayed from side to side.

SKEPTICISM AND HISTORY – DID THE HOLOCAUST HAPPEN?

There has been a good deal of argument among historians about exactly what historical knowledge consists of. This is a profound question, and we will not deal with it here. The main points are simply these. Historians have largely given up trying to propound general 'laws of history', corresponding to the laws of science. Instead, they concentrate on working out what happened in the past, and why it happened the way it did. Often, the attempts to work out reasons focus on historical figures' views of what was happening, and their reasons for acting as they did (Carr 1961). A key point about history is that it is specific: you will not find historians, in general, generalising across cultures or historical periods. If they do, then it is usually with great caution and many qualifications.

Of course, historians do make use of science in their work. They may use radioactive dating to work out the age of documents and artefacts. They may use biology to examine insects trapped in tombs, or x-ray and fluorescent techniques to examine pictures. Unlike scientists, however, historians are interested in particular explanations of particular events, rather than general laws and propositions.

As part of this research process, historians have built up an enormous body of knowledge about the past, about what happened and why. Since historical methods, in some respects, resemble those of science, the findings are usually considered reliable. There are criticisms and arguments in history, just as there are in science. Thus, when major historical findings are questioned by dissidents, this might be a useful place to see if skeptical techniques can work.

A brief survey of the literature shows that there are a large number of historical dissidents. There are people who believe that Francis Bacon wrote Shakespeare's plays (Gibson 1962). There are also people who argue that many gods and legendary heroes were actually kings and queens in Atlantis (Sladek 1978). And there is the view that human history has been heavily influenced by visiting aliens (von Däniken 1970). But perhaps the most controversial historical dissidents are the so-called Holocaust deniers, and we will examine them in some detail.

HOLOCAUST DENIAL

The most hideous cataclysm to strike the human race was undoubtedly the Second World War. About 50 million people died during the six years of horror, leaving entire countries smashed and partly depopulated (Bullock 1993). It is generally accepted that within this great conflagration another horror occurred, and this has come to be called the Holocaust.

Historians have researched the Holocaust fairly thoroughly – it is, after all, still just within living memory – and there is a broad consensus about what happened. The Nazis in Germany were strongly motivated by hatred of Jewish people, and cast about for a solution to the problem, as they saw it, of several million Jewish people in the parts of Europe under their control. They considered several ways of expelling the Jews, none of which were practicable. Finally they began to exterminate Jewish people and others of whom they disapproved (such as homosexuals and Gypsies). Many were starved or worked to death in concentration camps, and others died of disease in these dreadful conditions. As the war continued, the Nazis became more systematic in their efforts to exterminate the Jews. Many were shot and buried in mass graves, while large numbers were killed in purpose-built gas chambers. It is this latter claim that arouses especial horror. The Nazis, fighting desperately to avoid being overwhelmed in the war, diverted enormous resources to set up an industry of death in Europe.

Trains were chartered to transport tens of thousands of people to places such as Auschwitz and Treblinka. People were told that there they were to have showers, and were then herded naked into gas chambers. An extra dimension of horror is the lesser-known fact that the people did not die quickly. At the quickest, the Zyklon B gas that was used could kill all the people in about fifteen minutes (Shirer 1964, p. 1155), but sometimes it took hours.

To an outsider, the story of the Holocaust appears both horrific and stupid. It seems impossible that humans could set up an industry specifically for killing people, and then operate it relentlessly for years on end. Further, since Germany was at war against overwhelming enemies, it seems insane to devote resources to such a project.

Overwhelmingly, historians accept that the Holocaust happened. The evidence has been presented many times. One of the best, and most readable summaries is by the skeptic Michael Shermer. In essence, Shermer argues that a whole series of different types of evidence all point irresistibly to one conclusion: that the Nazis murdered about six million Jews (Shermer 1997).

What is the evidence? First, there are eyewitnesses. Both former guards and former inmates – including some who cleared bodies from the gas chambers – have testified that the Holocaust did occur. There are also written testimonies, such as an account of the killings written by a deceased victim and found beside a mass grave. Then there are artefacts which support the claims, such as warehouses at Auschwitz of eyeglasses, human hair and clothing. There is also a thick layer of human ash – metres thick – at Treblinka, hard to account for except in terms of hundreds of thousands of cremated bodies. Then there are records kept by the Germans, showing the transports with thousands of people arriving day after day at the death camps. Then there are records of meetings where Nazi leaders declared that the Jews were to be exterminated. And there are records which show the plans and contracts for the construction of gassing chambers. There is a great deal of evidence (Shermer 1997, pp. 214–24).

In addition to all this direct evidence, there is the work of demographers, which repeatedly shows that about six million Jewish people disappeared during the Second World War (Shermer 1997, p. 237). We would not be surprised if hundreds or thousands of people disappeared without trace during a major war, but six million is another matter. Anyone who wants to dispute the Holocaust has to account for all those missing people.

It seems clear that there is sufficient evidence for us to conclude that the Holocaust happened. There was a systematic attempt by the Nazis to wipe out the Jews, as well as millions of other people. Over the years, a mass of historical scholarship has added to this conclusion. The Holocaust is well documented, and the evidence amply meets any rational level of evidence required. Obviously, though, as with any evidence-based finding, enough contrary evidence could alter our view.

Faced with this mass of evidence, who would want to dispute it and why? We can say reasonably clearly who the disputants are. They are the people associated with the Institute for Historical Review, in the American state of California, and other similar groups. They include some noted – and some notorious – historians, and a good many other figures (Institute for Historical Review 2008).

What do the Holocaust deniers say? First, they argue that they do not 'deny' the Holocaust. Their key point is that far fewer people died than conventional historians claim, and there was no overall plan to kill the Jews. Instead, the deniers argue that the Germans, under Hitler, were entitled to treat the Jews with some fairly harsh measures. However, it is the fault of the allies – the British and the Americans – that many more people died. This is because the enormous allied bombing campaigns during the war prevented proper supply to the concentration camps, leaving the people there in atrocious conditions.

What of the various types of evidence we saw earlier? Deniers do not accept eyewitness evidence. The reports from the inmates of concentration camps are clearly unreliable, they say, as these people are predominantly Jewish. The confessions of concentration camp guards and other officials are also unreliable, as these were coerced from them after their capture. The documentation is all part of an enormous hoax started immediately after the war. So are the artefacts, the human hair and so on.

And what happened to all the people, the six million whom demographers claim to have been found missing? According to the deniers they were dispersed 'somewhere in the east', and their fate is, apparently, not a matter of great concern (Shermer & Grobman 2000).

It is apparent that there is a large element of conspiracy theory in these arguments. A constantly recurring theme is that 'the Jews' were greedy and exploitative and deserved some rough handling by Hitler and the Nazis. They are now securing unfairly good treatment for themselves by ensuring that they are revered because of the alleged horrors of the Holocaust. A recurring theme in the Holocaust deniers' literature is that 'the Jews should not receive any special treatment', and there is an underlying anger at the way in which the distress of Jewish people about the Holocaust is treated with such delicacy.

Holocaust deniers claim to be seeking the truth. Recurrent in their literature is the claim that all knowledge should be open to revision in the light of fresh evidence. For example the most prominent Australian Holocaust denier, Fredrick Töben, argues in these terms:

> Every thinking human being is a revisionist. Revisionism is nothing but a method, an heuristic principle, with which to construct one's world view. Opinions are constantly revised through a free flow of information. Only encrusted minds cannot absorb new information, preventing moral responsibility from coming to the fore. Then concerned citizens are arrested during private discussions and thrown into prison.
>
> (Töben 2006)

Apart from the last sentence, Töben's statement would accord well with a skeptical view of knowledge. In some circumstances even the last statement might also be true, such as in Hitler's Germany or Stalin's Russia. As we will see, the Holocaust deniers are engaged in a rather different activity. The nature of this activity has been the focus of at least two court cases, and in the normal course of events it would be argued and debated until a conclusion was reached. However, the Holocaust deniers did not take this route. Instead, a series of works appeared which did not analyse the responsibility for the war, but sought to re-write what happened in it.

Holocaust denial has its intellectual centre at the Institute for Historical Review in California. Despite its name, this is not affiliated to any academic organisation, such as a university. It is in fact in an unmarked building in an industrial park. There is a lock on the door and stringent security precautions (Shermer 1997, p. 191). The reason for all the security is simple: Holocaust deniers are not exactly popular, and there have been repeated examples of threats and harassment against them.

The Institute publishes assorted books, and also a journal called, logically, the *Journal for Historical Review*. It is easy to find back issues of this journal on the internet. It mimics an academic journal, and indeed Lipstadt (1993) tells of how a postgraduate student sent a paper to this journal, unaware of its notoriety.

In some countries, of course, Holocaust denial is in breach of the law. David Irving, the controversial historian, has been arrested in Germany on charges of denying that the Holocaust took place. He has been in prison in Austria for the same reason. The United States cannot forbid Holocaust denial, because of the constitutional amendment guaranteeing free speech, but the deniers have had their legal troubles there as well. One of the best known cases concerned Mel Mermelstein, an American of Jewish descent, whose father died in a concentration camp. Mermelstein wrote to a newspaper declaring that he could prove the Holocaust took place, and the Institute for Historical Review wrote challenging him to do so, and offering a reward of $50 000. Mermelstein took up the challenge and produced his evidence. When the Institute did not pay, he took them to court and won a judgement stating that he had proved the Holocaust did occur. He was awarded not $50 000 but $90 000, and it remains on the judicial record in the US that Mermelstein has proved that the Holocaust occurred (Shermer 1997, p. 191).

Another traumatic court case occurred in the UK, involving one of the most notorious of all Holocaust deniers, the English historian David Irving. Irving has no links to any university or academic institution: he supports himself through his books and speaking engagements. I have heard him on radio and television: he is a powerful, assertive debater who often seems to overwhelm his opponents. Initially, David Irving was not a Holocaust denier at all. He was a controversial writer whose best-known book was *Hitler's War* (Irving 1977). He argued not that the Holocaust did not occur, but that Hitler knew nothing about it. This caused great controversy, and seems highly unlikely, given that Hitler was notorious for his knowledge of even the minutest details of what was going on in the Reich.

Irving's London court case was set in motion when American academic Deborah Lipstadt published a book on Holocaust denial. Deborah Lipstadt is the head of a research centre at Emory University in the US, and a prominent commentator on Jewish affairs. In 1993 she published *Denying the Holocaust: The Growing Assault on Truth and Memory.* The book was a thorough analysis and condemnation of the growing Holocaust denial movement, backed by a great deal of research (Lipstadt 1993); it was published by Penguin in the UK. The book does not focus on Irving, though he is described as one of the most dangerous of the Holocaust deniers, perhaps half a dozen pages, in all, are devoted to him.

Irving sued Lipstadt and Penguin in the British courts, arguing that the book defamed him, lowered his standing in the eyes of those who might read Lipstadt's criticisms of him, and consequently damaged his chances of earning a living. He offered Lipstadt and Penguin an easy way to settle: donations to a charity, withdrawal of the book and a written apology. There was never any question of Lipstadt or Penguin accepting this offer, but the defendants faced a problem. There was little doubt that what the author had written about Irving was defamatory, and would cause him to be less well regarded. Therefore, they could not argue that they had not defamed him.

Among other matters, Lipstadt had characterised Irving in these terms: 'Irving ... has joined the ranks of the deniers' (Lipstadt 1993, p. 8). And, to make clear exactly what she thought about Holocaust deniers, she wrote that 'deniers misstate, misquote, falsify statistics and falsely attribute conclusions to reliable sources' (Lipstadt 1993, p. 111). And she also described Irving as a 'Hitler partisan wearing blinkers' and 'a self-described "moderate fascist"' (Lipstadt 1993, p. 161).

Lipstadt also made other criticisms, stating that Irving was an ardent admirer of the Nazi leader, and that he placed a self-portrait of Hitler over his desk, described his visit to Hitler's mountaintop retreat as a spiritual experience, and declared that Hitler reached out to help the Jews (Lipstadt 1993, p. 161).

This is painful criticism, and there is little wonder that Irving was greatly upset by it. That he chose to sue is more surprising, since he would be more aware than anyone of the truth or otherwise of the accusations. Still, he did sue in the UK, and chose to represent himself in court. Apparently he believed that since he knew the material better than anyone else, he was best placed to handle the case.

Lipstadt and Penguin faced a dilemma at this stage. A normal practice when being sued for defamation is to argue that the words are not defamatory. However, there is little doubt that anyone reading Lipstadt's account of Irving would think less of him, so this option was not available. Instead, the course of action the defendants chose was to argue that the statements were true. This is a surprising inversion of the normal procedure in court, and it follows from the basis of British law. Normally the onus is on the plaintiff or prosecutor to establish the case, but here the onus was on the defendants (Schneider 2001, p. 1532). Lipstadt and Penguin did not have to prove their case completely: they only had to prove that what had been written about Irving was substantially true, and this they set out to do.

Irving and the defendants had agreed that the case should be argued before a single judge rather than a jury, and Mr Justice Gray heard the case. The case ran for several months. Lipstadt did not give evidence, but heard the savage courtroom battle in its entirety. At the end, the judge concluded that Lipstadt and Penguin had met their burden of proof, and justified nearly all their claims. Part of his judgement is as follows:

> ... Irving has for his own ideological reasons persistently and deliberately misrepresented and manipulated historical evidence; that for the same reasons he has portrayed Hitler in an unwarrantedly favourable light, principally in relation to his attitude towards and responsibility for the treatment of the Jews; that he is an active Holocaust denier; that he is anti-Semitic and racist and that he associates with right wing extremists who promote neo-Nazism.
>
> Schneider (2001, p. 1533)

The judge therefore found in favour of Lipstadt and Penguin. He did not give them a completely clean sheet, finding for example that they had not established that Irving kept a picture of Hitler above his desk. However, the author and publisher had clearly won the case. Irving was left with a huge bill – for over three million pounds – which he could not meet. He went bankrupt and had to sell his house.

In both the Mermelstein and Lipstadt cases, therefore, the Holocaust deniers tried to argue their case in front of a professionally neutral judge. In both cases the onus was on the anti-denier (the deniers call them exterminationists) to establish their case, and in both cases they did so.

What should we make of Holocaust denial? The first point to be made is that we can usefully apply skeptical principles to the arguments. We do not know everything about the Holocaust, but there does seem to be a great deal of evidence that the Holocaust did happen. Therefore the burden of proof is now on the deniers: if they want to convince us that something different occurred, they must produce evidence comparable in scale and quality with what already exists. As both judges found, they have not managed to do this.

Occam's razor can also be applied. Clearly, something happened to Jewish people in Europe during the Second World War. We have much evidence from many sources about the extermination campaigns of the Nazis. The alternative hypothesis from the deniers appears to be that, yes, the Jews were interned in concentration camps that, yes, many of them died (through no fault of the Nazis, apparently) but that in addition to this, there was a gigantic conspiracy immediately after the war to fake documents, construct artefacts and formulate the stories of a giant imaginary crime. In short, the deniers seem to subscribe to a giant conspiracy theory, where the Jews are the major conspirators. And yet no details of this conspiracy have come to light: no documents have been found, no conscience-stricken participants have denounced the Holocaust as a fraud. In short, the deniers' theories are a good deal less economical than those of orthodox history.

Finally, the 'evidence' which the deniers have produced is not convincing. In the London court case, Lipstadt and Penguin had to show that Irving's evidence in his books was poor, and they were able to do this. One of the keys to the case was the evidence of Professor Richard Evans, a noted British historian. He was commissioned to examine Irving's work in detail, and worried Lipstadt because initially he seemed skeptical about her case. However, Evans's final conclusions, which he turned into a book, are quite devastating. In his book, he writes of how Irving 'falsely attributed conclusions to reliable sources, bending them to fit his arguments. He relied on material that turned out directly to contradict his arguments when it was checked.' (Evans 2001, p. 69)

> He quoted from sources in a manner that distorted their authors' meaning and purposes. He misrepresented data and skewed documents. He used insignificant and sometimes implausible pieces of evidence to dismiss more substantial evidence that did not support his thesis. He ignored or deliberately suppressed material when it ran counter to his arguments. When he was unable to do this, he expressed implausible doubts about its reliability.
>
> (Evans 2001, pp. 69–70)

It is important to notice that Evans is not saying just that Irving made many mistakes. He is also saying that Irving's mistakes systematically distort the historical record, so that a misleading picture of the past emerges. This point was made clear in the court case, when Evans was cross-examined by Irving about the latter's use of documents.

> MR. IRVING: Do you say that I misinterpreted and distorted them deliberately? Is that your contention?

> PROF. EVANS: Yes, that is my contention. You know there is a difference between, as it were, negligence, which is random in its effects, i.e. if you are simply a sloppy or bad historian, the mistakes you make will be all over the place. They will not actually support any particular point of view . . . On the other hand, if all of the mistakes are in the same direction in the support of a particular thesis, then I do not think that is mere negligence. I think that is a deliberate manipulation and deception.
>
> (Evans 2001, p. 205)

In short, expert examination of Irving's work showed it to be profoundly misleading. Irving's use of evidence was unreliable, and systematically distorted in favour of his basic thesis. Since the deniers are making an extraordinary claim, the evidence clearly fails to meet this standard, and indeed should not convince anyone at all. In short, it resoundingly fails to tip Sagan's balance.

Why do people deny the Holocaust? In some cases the reasons are very simple. Anti-Semitism still exists in the world, and anti-Semites find it infuriating that the story of the Holocaust – the greatest of all crimes against the Jews – is widely accepted. It is not difficult to verify this. If one examines a website related to Holocaust denial, one usually finds that it starts off with high-minded sentiments about seeking the truth and being open to contradictory evidence. However, usually, a few links away, it is easy to find anti-Semitic cartoons or comments, or simply repeated references to 'the Jews'. The assumption of deniers appears to be that all Jews see things in exactly the same way, and all are part of the conspiracy.

In addition, as Michael Shermer points out, the deniers seem to have a particular mindset which propels them:

> Many deniers seem to like the idea of a rigid, controlled and powerful state. Some are fascinated with Nazism as a social/political organization and are impressed with the economic gains Germany made in the 1930s and her military gains from 1939 to 1941. The history of the Holocaust is a black eye for Nazism. Deny the veracity of the Holocaust, and Nazism begins to lose this stigma.
>
> (Shermer 1997, p. 252)

Shermer also points out that there are similarities between Holocaust denial and creation science. In both cases there is a massive process, generally accepted, for which the evidence is overwhelming. In both cases there are people with strong motives for not accepting the process. In both cases, the dissenters conduct a kind of 'guerrilla war' against the predominant belief. Shermer summarises the tactics as follows:

> 1. Holocaust deniers find errors in the scholarship of historians and then imply that therefore their conclusions are wrong, as if historians never made mistakes. Evolution deniers (a more appropriate title than creationists) find errors in science and imply that all of science is wrong, as if scientists never make mistakes.

> 2. Holocaust deniers are fond of quoting, usually out of context, leading Nazis, Jews and Holocaust scholars to make it sound like they are supporting Holocaust deniers' claims. Evolution deniers are fond of quoting leading scientists like Stephen Jay Gould and Ernst Mayr out of context and implying that they are cagily denying the reality of evolution.

> 3. Holocaust deniers contend that genuine and honest debate between Holocaust scholars means that they themselves doubt the Holocaust or cannot get their stories straight. Evolution deniers argue that genuine and honest debate between scientists means even they doubt evolution or cannot get their science straight.
>
> (Shermer 1997, p. 132)

This is a dazzling insight. It suggests that a new form of political movement is coming into being. This is a pseudoacademic wing of social movements. People with a strong belief masquerade as academics asking for free and open debate on some topic, while being strongly committed to a particular view in advance.

In sum, the skeptical principles do help us make sense of the Holocaust denial debate. The terms burden of proof, Occam's razor and Sagan's balance are applicable, and are surprisingly analogous to what might be encountered in the debate over creation science. This is because historical knowledge, at its best, can be comparable to scientific knowledge in reliability.

EVIDENCE-BASED MEDICINE

One impression may be troubling the reader by this stage. It may look as if the skeptically based approaches are terribly authoritarian in their effects; it is always the outsiders, the non-conformists and the people willing to take unorthodox approaches whom skeptics criticise. To some degree this is so. It does seem as if many of the 'way-out' therapies and approaches have little to recommend them in the way of evidence. However, there is one area where skepticism, or one of its related approaches, is having a radical effect, questioning accepted judgements and changing established ways of thought and action. This is the area of evidence-based medicine (EBM).

To understand the impact of evidence-based medicine, we have to understand the medical profession a bit better. Medicine has always been involved with areas of human concern – life, death, pain and health – where people may be desperate and terrified. After all, life and death are the routine concerns of the medical profession. However, the actual ability of doctors to deliver cures has always been limited. There is still a great deal that doctors do not know, and many diseases where it is not clear what the best treatment is.

In response to this, doctors have traditionally been given a great deal of autonomy. That is, they have a large degree of discretion in deciding what treatment should be given to the patient. Gradually, this situation has begun to change. As we have seen, the double-blind controlled trial is the key standard for deciding what treatments work and what do not. As we have seen with Dr Stephen Bratman's experience, it is perfectly possible for doctors to see patients visibly improving under their care while they are being treated in ways that are completely useless. The double-blind controlled trial is vital for distinguishing what treatments work and what do not.

Large numbers of double-blind controlled trials were carried out in the decades of the twentieth century. If they had been left to moulder in the literature, this would have been a waste of time and effort. It was an outstanding British doctor, Archie Cochrane, who set out to remedy this situation. His most important contribution to medicine came in 1972 with a book on the British health service titled *Effectiveness and Efficiency – Random Reflections on Health Services* (Cochrane 1972), where he argued that it was essential to target limited resources in order to deliver health as efficiently as possible. This is, of course, a fairly simple idea. What made Cochrane unusual was that he also suggested that one way of using the resources efficiently was to build up a database of information about which medical treatments worked and which did not. Further, he suggested that this database should use the results of randomised controlled trials to ensure that the treatments were the best. It may come as a shock to many non-medical people that such information was not already readily available to doctors.

Cochrane's ideas were taken up, and are now enshrined in the Cochrane Collaboration, a worldwide network of doctors centred in Oxford, UK. The Collaboration now has over 50 collaborative groups and publishes reviews on the effectiveness of treatments in different areas, using the best evidence available. Gradually, this evidence-based movement is transforming medicine.

Let us look at an example of how evidence-based medicine works. Here is a case from an American book on the topic: it appears in an attached CD.

> A middle aged woman presented with exacerbation of her usual pattern of migraine. She asked specifically whether riboflavin might help as she had heard from a friend that it helped prevent migraine. Together you formulate the question: in patients with frequent migraines, is riboflavin effective in the reduction of migraine frequency or severity?
>
> (Sackett et al 2000)

Note the care that has to be taken in formulating the question: a poorly thought-out question can only lead to misleading answers. The doctor in question then went to the Cochrane databases and looked for answers to the question. The doctor found a research paper which did seem to answer the question. It was published in the journal *Neurology* in 1998. The study seemed to be well constructed, and came to these conclusions:

> Treating 2 patients with migraine with 400mg/day riboflavin will result in one of them having a 50% reduction in migraine frequency (e.g. 4–2 per mo.)

> Caveats: the dose takes about 3 months to take effect, costs about $10US per month, and the study size is rather small.
>
> (Sackett et al 2000)

This result can then be taken back to the patient, who can be told that if she takes the prescribed dose, there is a 50 per cent chance that she will have a 50 per cent reduction in the frequency of her migraines. I know people who suffer from migraine. My guess would be that virtually all of them, faced with the misery of a life racked by recurrent migraines, would be only too happy to try the riboflavin treatment. It looks well worth the effort.

Let us look at another case, where the results are not so clear-cut. This concerns a child's hyperactivity. I have known parents with children like this, and it can drive people to the edge of despair. Let us see what happens in this case.

> A boy has ADD with hyperactivity. The mother has found leaflets from a parents' group which suggest that hyperactivity can be caused by "hidden allergies" to common foods. They recommend "few foods, low sugar, diet". You pose the question: "In children with hyperactivity, does any form of dietary modification improve behaviour?"
>
> (Sackett et al 2000)

Note again the care with which the question is posed. We might also remember that, in practice, we might find that the question might have to be re-phrased. There might be no answers to a narrow question, and far too many answers to a broad one. How does evidence-based medicine work in this case? The first thing the search revealed was a vast number of irrelevant papers. The mother in the story especially wanted to know about sugar, but the researcher found a study which indicated that sugar had no effect. Eventually, after a good deal of searching, two papers were located which seemed to bear directly on the topic of diet and children's behaviour. Alas, one of them was in German, and so was discarded from the study. (In a perfect world, of course, it would be possible to find someone who could translate the paper, or at least summarise it in English.) The other paper was also directly relevant (Schmidt et al 1997). It was a double-blind study of children admitted to a psychiatric ward. They had hyperactivity, and most had ADD as well. The children's diet was controlled to exclude items such as cereals. It seemed a good quality study, and the results could probably be accepted with confidence.

What was the bottom line? It was that an oligoantigenic diet (i.e. one which is believed least likely to cause allergic reactions) improves behaviour scores by more than 25 per cent in one in five children with ADD and/or conduct disorder.

We should note that this result is quite different from the previous one. There, there was a 50 per cent chance of a 50 per cent reduction in migraines. Here, there is only a 20 per cent chance of a 25 per cent improvement in the child's behaviour. Still, it is a result that can be brought to the mother, and a decision could be made about what to do. Would the mother decide to control her son's diet in the way suggested? That is up to her: the chances are not very good, and the improvements not very great, but she may decide it is worth a try.

What should we make of evidence-based medicine? First, as Bratman (2005, p. 64) points out, it is revolutionising medicine. For the first time, doctors can prescribe treatments which they have reason to believe are the best available. What is more, they can give patients a good idea of the probability that the treatment will work, and even how much of an improvement is likely.

On the other hand the results of evidence-based medicine are always in probabilistic form: there is a certain percentage chance of a certain percentage improvement. The doctor can no longer adopt a confident manner to cheer the patient up: the facts are before both. In addition, the evidence-based medicine procedures are often slow, and the patients may sometimes not be able to grasp exactly what the results mean.

It is likely that evidence-based medicine will become more and more prominent in medical practice. After all, it would only take a few court cases where patients suffered because doctors or surgeons had not performed the best evidence-based treatment to convince all doctors that this is the way to go.

It is also true that evidence-based medicine has its limits. Since it is relatively slow, doctors in casualty and emergency wards probably cannot ever apply the methods in their entirety. They have to make quick decisions, which do not allow for prolonged consideration of evidence. Again, some people – especially elderly patients – often have complex problems where the evidence-based approaches give no guidance. Finally, sometimes a treatment is so obviously beneficial that there is no need to look up evidence: one simply goes ahead and treats. Thus evidence-based medicine is of great value, but it is not a complete panacea.

However, the impact of evidence-based methods is not likely to be confined to medicine, or to the other medically related professions such as nursing, chiropody and so on. The methods are equally applicable to teaching – at all levels from primary to tertiary – to social work interventions, financial planners and forecasters and much more (e.g. Thomas and Pring 2004).

Not all of these evidence-based interventions will be welcomed. Ayres (2007) has described the range and impact of evidential approaches, and points out that they acquire extra power because of the advent of high-powered computing, which enables a huge range of inferences to be drawn from data. In some cases this is extremely intrusive. For example Ayres describes the advent of 'direct instruction'. In this system, based on a mass of evidence, teachers are not only taught what to teach: their every act is precisely scripted, down to where they point and the questions they ask the students (Ayers 2007, p. 156–62). This would be regarded as intolerable, except that the evidence shows that the required approaches lead to better educational outcomes. There are clearly great arguments ahead in deciding exactly how far evidentialism can go.

As we have seen, skepticism is not something that stands alone. In my view, it is a part of a larger movement of evidence-based approaches. These approaches started in science, with people such as Galileo insisting that evidence must determine scientific theories. Slowly, over several centuries, they spread through science and began to influence science-based technologies such as medicine. Skeptics also used them in the 20th century to evaluate paranormal claims. In my view, they can now be expanded to a whole range of other areas. As with evidence-based medicine, they do not solve all the problems, but they do improve decisions and understanding in a large number of cases.

LAUNCHING PAD

In this chapter we have gone beyond the investigation of claims of the paranormal to ask if there are claims of non-paranormal knowledge which skepticism can investigate. Strictly, by the definitions we have applied, the answer is no: since skepticism is the investigation of the paranormal, it cannot apply to other areas. It does seem clear that something very similar to skeptical approaches is useful and we have seen how it can usefully be applied to the Holocaust denial, to claims that the *Apollo* Moon mission was a hoax, and also to urban myths of the kind investigated by the *MythBusters*. Probably the major difference between the paranormal claim and these other claims lies in the application of Sagan's balance. The presupposition of falsity is usually less strong for non-paranormal claims, and so less evidence is required to substantiate them.

Moving beyond dissenting claims of knowledge, we also saw that there is a substantial movement in medicine towards evidence-based medicine, in which the results of randomised clinical trials are applied to doctors' actions. Gradually, it looks as if medicine will be transformed by these changes. More controversially, evidence-based approaches are being applied to other areas of practice, such as teaching, although the results are far less dramatic.

What this suggests is that skepticism is not a stand-alone phenomenon. It is part of an evidence-based movement that is beginning to exert a substantial effect on all parts of society. It is likely that this effect will be greatly increased in the future.

This book has been an argument for skepticism. It has been suggested that skepticism can be boiled down to three basic principles and that these – in conjunction with knowledge about particular claims – can enable anyone to make good judgements about what to believe. In addition, it has been argued that skepticism, when applied widely, would make the world a better place. Some paranormal beliefs are damaging and dangerous, and skeptical arguments are one way of keeping them in check. There are strong ethical reasons for adopting skepticism as a general approach. Finally, skepticism can be the most enormous fun. Enjoy it!

Shermer's long (Shermer & Grobman 1997) and short (Shermer 1997) works on the Holocaust deniers are both good, straightforward introductions, and well argued cases against these people. For the work of a specialist historian, checking the fine grain of denial claims, Evan's book (2001) shows how he analysed David Irving's claims. It also gives an account of his days in the witness box during the court case. Lipstadt (2005) has written her own account of this. Of course, the Institute for Historical Review (2006) is always anxious to put forward its own case, and this can be found on its website.

Philip Plait's debunking of *Apollo* hoax myths is entertaining, and his book also sets about debunking other astronomy-related myths (Plait 2002). As we have seen, evidence-based medicine is an increasingly important area, and an introduction such as that of Sackett and his colleagues (Sackett et al 1997) is well worth the effort. A paper, recommended earlier, by Bratman (2005) gives a good idea of the impact that the evidence-based movement is having on medical work. For an idea of the controversy in other areas, the speech by Hargreaves (1996) advocating evidence-based teaching, and the resulting controversy between Hammersley (1997) and Hargreaves (1997) is worth a look. The book by Ayres (2007) shows the astonishing potential of evidence-based approaches, though some of them are clearly controversial.

Bibliography

Abell, GO 1981, 'Astrology', in George O Abell and Barry Singer (eds), *Science and the paranormal: probing the existence of the supernatural*, Charles Scribner's Sons, New York, pp. 70–94.

Aldred, L 2002, 'Money is just spiritual energy', *Journal of Popular Culture*, vol. 35, no. 4, pp. 61–74.

Anonymous 1999, 'Breatharians breathe their last', *Skeptic*, vol. 7, no. 4, p. 28.

—— 2000, 'The ten outstanding skeptics of the twentieth century', *Skeptical Inquirer*, vol. 24, no. 1, pp. 23–8.

Australian Skeptics 2003, *The Great Water Divining DVD*, Australian Skeptics, Roseville, NSW.

—— 2007, 'Costly advice', *The Skeptic*, vol. 27, no. 1, p. 6.

—— 2008, <http://www.skeptics.com.au> (accessed 30 December 2008).

Ayres, I 2007, *Super crunchers. How anything can be predicted*. John Murray, London.

Bacon, F 1625, 'Of prophecies', in Francis Bacon, *The essays or counsels, civic and moral, of Francis Ld. Verulam, Viscount of St. Albans*, Champaign, Illinois, Project Gutenberg, pp. 43–4.

—— 1856 [1620], 'Novum organum, or, the suggestions for the interpretation of nature', in Joseph Devey (ed), *The physical and metaphysical works of Lord Bacon*, Henry G. Bohm, London, pp. 380–567.

Bagger, MC 2002, 'The ethics of belief: Descartes and the Augustinian tradition', *The Journal of Religion*, vol. 82, no. 2, pp. 205–25.

Barr, J 1984, *Escaping from Fundamentalism*, SCM Press, London.

Bartholemew, RE, Goode E 2000, 'Mass delusions and hysterias: highlights from the past millennium', *Skeptical Inquirer*, vol. 24, no. 3, pp. 20–5.

Bartz, WR 2002, 'Teaching skepticism via the critic acronym and the Skeptical Inquirer', *Skeptical Inquirer*, vol. 26, no. 5, pp. 42–4.

Becker, J 1996, *Hungry ghosts: China's secret famine*, John Murray, London.

Beecher, HK 1955, 'The powerful placebo', *Journal of the American Medical Association*, vol. 159, pp. 1602–6.

Bentham, J 1789, *An introduction to the principles of morals and legislation*, T Payne and Son, London.

Blackmore, S 1996, *In search of the light. The adventures of a parapsychologist*. Prometheus Books, Amherst, New York.

Blackmore, S, Troscianko, T 1985, 'Belief in the paranormal: probability judgements, illusory control and the "chance baseline shift"', *British Journal of Psychology*, 76, pp. 459–69.

Bliss, RB 1976, *Origins: two models – evolution and creation*, Creation-Life Publishers, San Diego.

Blum, D 2007, *Ghost hunters. The Victorians and the hunt for proof of life after death*, Century, London.

Bok, BJ et al 1975, 'Objections to astrology', *The Humanist*, September/October, p. 6.

Bratman, S 2005, 'The double-blind gaze', *Skeptic* (Altadena USA), vol. 11, no. 3, pp. 64–73.

—— 2006, 'Who is Steven Bratman?', <http://www.orthorexia.com/index.php?page= Bratman> (accessed 22 March 2006).

Braude, S 1979, *ESP and psychokinesis: a philosophical examination*, Temple University Press, Philadelphia.

Bridgstock, M 1978, *Do-it-yourself pseud detector*, Griffith University, Brisbane.

—— 1986a, 'The reliability of creationist claims', in Martin Bridgstock and Ken Smith (eds), *Creationism: an Australian perspective*, Australian Skeptics, pp. 63–6.

—— 1986b, 'What is the Creation Science Foundation Ltd?', in Martin Bridgstock and Ken Smith (eds), *Creationism: An Australian perspective*, Australian Skeptics, pp. 81–5.

—— 1989, 'The twilit fringe – anthropology and modern horror fiction', *Journal of Popular Culture*, vol. 23, no. 3, pp. 115–23.

—— 1998a, 'The scientific community', in Martin Bridgstock, David Burch, John Forge, John Laurent and Ian Lowe, *Science, technology and society: an introduction*, Cambridge University Press, Melbourne, pp. 15–39.

—— 1998b, 'Controversies regarding science and technology', in Martin Bridgstock, David Burch, John Forge, John Laurent and Ian Lowe, *Science, technology and society: an introduction*, Cambridge University Press, Melbourne, pp. 83–107.

—— 2003, 'Paranormal beliefs among science students', *Skeptic*, vol. 23, no. 1, pp. 6–10.

—— 2004, 'Teaching skepticism at the university', *Skeptic*, vol. 24, no. 3, pp. 12–4.

—— 2008, 'Skeptical ethics – what should we investigate', *Skeptical Inquirer*, vol. 32, no. 3, pp. 35–9.

Bridgstock, M, Burch, D, Forge J, Laurent, J and Lowe, I 1998a, *Science, technology and society: an introduction*, Cambridge University Press, Melbourne.

—— 1998b, 'Introduction', in Martin Bridgstock et al *Science, technology and society: an introduction*, Cambridge University Press, Melbourne, pp. 3–14.

Bridgstock, M, Smith, K (eds) 1986, *Creationism: an Australian perspective*, Mark Plummer for the Australian Skeptics, Melbourne.

Broad, W, Wade, N 1985, *Betrayers of the truth*, Oxford University Press, Oxford and New York.

Brugger, P, Landis, T, Regard, M 1990, 'A sheep-goat effect in repetition avoidance', *British Journal of Psychology*, 81, pp. 455–68.

Brugger, P, Taylor, KI 2003, 'ESP – Extrasensory perception or effect of subjective proba-bility?', *Journal of Consciousness Studies*, vol. 10, nos. 6–7, pp. 221–46.

Buchhorn, R 1999, 'Cannibalism lives', *Skeptic*, vol. 19, no. 4, pp. 51–4.

Buechler, SM 2000, *Social movements in advanced capitalism*, Oxford University Press, New York.

Bugliosi, V 1999, *Helter skelter*, Bantam, New York.

Bullock, A 1993, *Hitler and Stalin. Parallel lives*, Fontana, London.

Burns, RM 1981, *The great debate on miracles: from Joseph Glanvill to David Hume*, Bucknell University Press, Lewisburg.

Calaprice, A (ed) 1996, *The quotable Einstein*, Princeton University Press, Princeton.

Carey, TV 2005, 'The ontological argument and the sin of hubris', *Philosophy Now*, no. 53, pp. 24–7.

Carr, EH 1961, *What is history?*, Macmillan, London.

Carroll, RT 2008, *The skeptics' dictionary*, <http://www.skepdic.com> (accessed 20 December 2008).

Caso, A 2002, 'Three skeptics' debate tools examined', *Skeptical Inquirer*, vol. 26, no. 1, pp. 37–41.

Castle, E, Thiering, B 1973, *Some trust in chariots*, Westbooks, Perth and Sydney.

Chalmers, A 1988, *What is this thing called science?*, University of Queensland Press, St. Lucia.

Charpak, G, Broch H 2004, *Debunked! ESP, telekinesis, and other pseudoscience*, Bart K Holland (trans.), Johns Hopkins University Press, Baltimore.

Clancy, SA 2005, *Abducted: how people come to believe they were kidnapped by aliens*, Harvard University Press, Cambridge, Mass.

Clark, M 1997, *Reason to believe*, Avon, New York.

Clarke, AC 1962, *Profiles of the future*, Pan, London.

Clifford, WK 1879, 'The ethics of belief', in Leslie Stephen and Frederick Pollock (eds), *William Kingdom Clifford: lectures and essays*, Macmillan, London, pp. 177–211.

Close, F 1992, *Too hot to handle*, Penguin, London.

Cochrane, AL 1972, *Effectiveness and efficiency: random reflections on health services*, Nuffield Provincial Hospitals Trust, London.

Cole, S 1992, *Making science. Between nature and society*, Harvard University Press, Cambridge, Mass.

Collins, HM 1975, 'The seven sexes: a study in the sociology of a phenomenon, or the replication of experiments in physics', *Sociology*, 9, pp. 205–24.

Collins, HM 1981, 'Son of seven sexes: the social destruction of a physical phenomenon', *Social Studies of Science*, 11, pp. 33–62.

Committee for the Scientific Investigation of Claims of the Paranormal, 2006, <http://www.csicop.org> (accessed 1 June 2006).

Crick, FHC, Watson, JD 1953, 'Molecular structure of nucleic acids', *Nature*, 171, pp. 737–8.

Curtain, J, Newbrook, M 1998, 'Oates' theory of reverse speech: an update'. *Skeptic*, vol. 18, no. 4, pp. 15–6.

Darwin, C 1859, *On the origin of the species by means of natural selection, or, the preservation of favoured races in the struggle for life*, John Murray, London.

Davis, A 2005, 'Psychic swindlers', *Skeptical Inquirer*, vol. 29, no. 3, pp. 38–42.

Degenhardt, MAB 1998, 'The ethics of belief and the ethics of teaching', *Journal of Philosophy of Education*, vol. 32, no. 3, pp. 333–44.

Descartes, R 1911 [1641], *Meditations on first philosophy*, ES Haldane (trans), Cambridge University Press, Cambridge.

Diaconis, P, Mosteller, F 1989, 'Methods for studying coincidences', *Journal of the American Statistical Association*, vol. 84, no. 408, pp. 853–61.

Edwards, H, Stollznow, K 1998, '"Alternative" consultations', *Skeptic*, vol. 18, no. 2, pp. 10–4.

Evans, RJ 2001, *Lying about Hitler*, Basic Books, New York.

Festinger, L, Riecken, HW, Schachter, S 1950, *When prophecy fails*, University of Minnesota Press, Minneapolis.

Fiorello, CA 2001, 'Common myths of children's behavior', *Skeptical Inquirer*, vol. 25, no. 3, pp. 37–44.

Fox, K 2004, *Watching the English*, Hodder and Stoughton, London.

Frazier, K 2001, 'From the editor's seat: 25 years of science and skepticism', *Skeptical Inquirer*, vol. 25, no. 3, pp. 46–9.

Frazier, K and the Executive Council 2007, 'It's CSI now', *Skeptical Inquirer*, vol. 31, no. 1, pp. 5–6.

French, CC, Richards, A 1993, 'Clock this! An everyday example of a schema-driven error in memory', *British Journal of Psychology*, 84, pp. 249–53.

Gardner, M 1957, *Fads and fallacies in the name of science*, Dover, New York.

Gibbon, E 1879 [1782], *The history of the decline and fall of the Roman Empire*. 6 vols, vol. 3, Harper & Brothers, New York.

Gibson, HN 1962, *The Shakespeare claimants: a critical survey of the four principal theories concerning the authorship of the Shakespearean plays*, Methuen, London.

Goldberg, ED 1965, 'Minor elements in seawater', in JP Riley and G Skirrow (eds), *Chemical Oceanography*, pp. 163–96.

Good, T 1987, *Above top secret: the worldwide UFO conspiracy*, Sidgwick and Jackson, London.

Goode, E 2000, *Paranormal beliefs: a sociological introduction*, Waveland Press, Prospect Heights, Ill.

—— 2008, 'The skeptic meets the moral panic', *Skeptical Inquirer*, vol. 32, no. 6, pp. 37–41.

Gorst, M 2001, *Aeons*, Fourth Estate, London.

Greenberg, D 2005, 'Flashbulb memories', *Skeptic* (Altadena USA), vol. 11, no. 3, pp. 74–80.

Grene, M 1999, 'Descartes and skepticism', *The Review of Metaphysics*, vol. 52, no. 3, 553–66.

Groarke, L 1990, *Greek Scepticism*, Queen's University Press, Montreal.

Guardian 2002, Adventist Sect Parents Jailed for Manslaughter, 14 June 2002, <http://www.guardian.co.uk/world/2002/jun/14/childprotection.uk> (accessed 29 June 2009).

Hagstrom, WO 1965. *The scientific community*, Basic Books, New York.

Hammersley, M 1997, 'Educational research and teaching: a response to David Hargreaves' TTA Lecture', *British Educational Research Journal*, vol. 23, no. 2, pp. 141–61.

Hanley, C, Mummery, K 2008, *Queensland Social Survey 2008. Final Sampling Report*, Central Queensland University, Rockhampton.

Hansen, GP 1992, 'CSICOP and the skeptics: an overview', *Journal of the American Society for Psychical Research*, vol. 86, no. 1, pp. 19–63.

Hardy, A, Harvie, R, Koestler, A 1973, *The challenge of chance*, Hutchinson, London.

Hargreaves, D 1996, *Teaching as a research-based profession: possibilities and prospects*, Teacher Training Agency, London.

Hargreaves, DH 1997, 'In defence of research for evidence-based teaching: a rejoinder to Martyn Hammersley', *British Educational Research Journal*, vol. 23, no. 4, pp. 405–19.

Hassan, S 2000, *Empowering people to think for themselves*, Freedom of Mind Press, Somerville, MA.

Hecht, JM 2003, *Doubt: a history*, HarperCollins, New York.

Hempstead, M 1992, 'All you need to know about crop circles', *Skeptic*, vol. 12, no. 2, pp. 12–9.

Hills, B 2000, 'Cheating death', *Sydney Morning Herald*, 30 December 2000, p. 25.

Hodgson, R, Davey, SJ 1987, 'The possibilities of mal-observation and lapse of memory from a practical point of view', *Proceedings of the Society for Psychical Research*, 4, pp. 381–495.

Hoffman, GA, Harrington, A, Fields, HL 2005, 'Pain and the placebo: what have we learned?', *Perspectives in Biology and Medicine*, vol. 48, no. 2, 248–65.

Holden, D 2006, 'The need for comprehensive naturopathic oncology clinics in NZ', <http://www.osiris.org.nz/Oncology/Oncology_NeedForOncologyClinics.htm> (accessed 10 January 2006).

Houdini, H 1980, *Miracle mongers and their methods*, Coles, Toronto.

Hume, D 1762, *The history of England*, 2 vols, A Millar, London, 1786.

—— 1964 [1739–40], 'A treatise of human nature', in TH Green and TH Grose (eds), *David Hume. The Philosophical Works*, vols. 182, Scientia Verlag Aalen, London.

—— 1980 [1748], 'Of miracles', in TH Green and TH Grose (eds), *David Hume. The Philosophical Works*, vol. 4, Scientia Verlag Aalen, London.

Hyde, V 2001, 'New Zealand Tragedy', *Skeptic*, vol. 21, no. 3, pp. 12–4.

Institute for Historical Review 2008, <http:// www.ihr.org> (accessed 3 January 2009).

Irving, D 1977, *Hitler's war*, Hodder & Stoughton, London.

Irwin, HJ 1989, 'On paranormal disbelief: the psychology of the skeptic', in GK Zollschan, JK Schumaker and GF Walsh (eds), *Exploring the Paranormal*, Prismunity, UK and Australia, pp. 305–12.

Jacobs, DM 1998, *The threat*, Fireside, New York.

James, W 1950 (1897), *The will to believe*, Dover, New York.

Jung, CG 1960, *The structure and dynamics of the psyche*, T. Hull (trans), London: Routledge and Kegan Paul, London.

Jungk, R 1965, *Brighter than a thousand suns*, Penguin, Harmondsworth.

Kahneman, D, Tversky, A 1979, 'Prospect theory: an analysis of decision under risk', *Econometrica*, 47, pp. 263–91.

Kanigel, R 1991, *The man who knew infinity. A life of the genius Ramanujan*, Charles Scribner's Sons, New York.

Kawa, S 2003, 'How I reluctantly became a skeptic, *Skeptic* (Altadena USA), vol. 10, no. 2, pp. 29–30.

Keay, C 2001a, 'Nuclear myths. The dirty thirty. Pt I', *Skeptic*, vol. 21, no. 1, pp. 53–6.

—— 2001b, 'The dirty thirty. Pt I', *Skeptic*, vol. 21, no. 2, pp. 45–8.

Keene, ML 1997, *The psychic mafia*. Prometheus Books, Amherst, New York.

Keller, EF 1983, *A feeling for the organism: the life and work of Barbara McClintock*, W.H. Freeman, San Francisco.

Kelly, L 2004, *The skeptic's guide to the paranormal*, Allen & Unwin, Crows Nest NSW.

Kennedy, D 2004, 'The old file-drawer problem', editorial, *Science Forum*, vol. 305, no. 5683, pp. 451.

Kitcher, P 1982, *Abusing science: the case against creationism*, MIT Press, Cambridge, Mass. and London.

Klass, PJ 1974, *UFOs explained*, Vintage, New York.

—— 1976, 'UFOs: fact or fantasy', *Humanist*, pp. 9–13.

—— 1983, *UFOs: the public deceived*, Prometheus Books, Amherst, New York.

Knight, J 1986, '"Creation-science" in Queensland: some fundamental assumptions', *Social Alternatives*, vol. 5, no. 3, pp. 26–31.

Koepsell, D 2006, 'The ethics of investigation', *Skeptical Inquirer*, vol. 30, no. 1, pp. 47–50.

Koestler, A 1969, *Arrow in the blue*, Hutchinson, London.

—— 1974, *The roots of coincidence*, Hutchinson, London.

Koyre, A 1965, *Newtonian studies*, University of Chicago Press, Chicago.

Kuhn, TS 1957, *The Copernican revolution*, Harvard University Press, Cambridge, Mass.

Kuhn, TS 1966, *The structure of scientific revolutions*. University of Chicago Press, Chicago.

Kurtz, P 1986, *The transcendental temptation*, Prometheus, Buffalo, New York.

—— 1992, *The new skepticism*, Prometheus, Buffalo, New York.

—— 1998, 'The new skepticism – a worldwide movement', *Skeptic*, vol. 18, no. 2, pp. 23–8.

—— 2001a, *Skeptical Odysseys*, Prometheus Books, Amherst, NY.

—— 2001b, 'A quarter century of skeptical inquiry', *Skeptical Inquirer*, vol. 25, no. 4, pp. 42–7.

—— 2007, 'New directions for skeptical inquiry', *Skeptical Inquirer*, vol. 31, no. 1, p. 7.

Le Grand, HE 1990, *Drifting continents and shifting theories*, Cambridge University Press, Cambridge.

Lead, R 1998, 'Bad scams, good scams, better scams', *Skeptic*, vol. 18, no. 4, pp. 17–20.

Lipstadt, DE 1993, *Denying the Holocaust: the growing assault on truth and memory*, Free Press, New York.

—— 2005, *History on trial: my day in court with Holocaust denier David Irving*, Ecco, New York.

Loftus, EF 1979, *Eyewitness testimony*, Harvard University Press, Cambridge, Mass and London, England.

—— 1997, 'Creating false memories', *Scientific American*, vol. 277, no. 3, pp. 71–5.

Loftus, EF, Miller, DG, Burns, HJ 1978, 'Semantic integration of verbal information into a visual memory', *Journal of Experimental Psychology: Human Learning and Memory*, 4, pp. 19–31.

Loftus, EF, Palmer, JC 1974, 'Reconstruction of automobile destruction: an example of the interaction between language and memory', *Journal of Verbal Learning and Verbal Behavior*, 13, pp. 585–9.

Lowe, I 2002, 'Neither skeptical nor an environmentalist', *Skeptic*, vol. 22, no. 4, pp. 61–3.

Loxton, D 2008, 'Where do we go from here?', *Skeptical Briefs*, vol. 18, no. 1, pp. 1–3.

Lyell, C 1830–1833, *Principles of geology: being an attempt to explain the former changes of the Earth's surface, by reference to causes now in operation*, J. Murray, London.

Lynn, J, Jay, A 1983, *Yes Minister*, British Broadcasting Corporation, London.

Mackay, J et al (eds) 1984, *The quote book*, Creation Science Foundation, Brisbane.

MacLennan, AH, Myers, SP, Taylor, AW 2006, 'The continuing use of complementary and alternative medicine in South Australia: costs and beliefs in 2004', *Medical Journal of Australia*, vol. 184, no. 1, pp. 27–31.

Magill, FN (ed) 1990, *The Nobel Prize winners*, vol. 3, Salem Press, Pasadena, California.

Marks, D 1981, 'Sensory cues invalidate remote viewing experiments', *Nature*, 292, p. 177.

—— 2000, *The psychology of the psychic*, 2nd edn, Prometheus, Buffalo, New York.

Marks, D, Kammann R 1978, 'Information transmission in remote viewing experiments', *Nature*, 274, pp. 680–1.

—— 1980, *The Psychology of the Psychic*, Prometheus, Buffalo, New York.

Marks, D, Scott, C 1986, 'Remote viewing exposed', *Nature*, 319, p. 444.

McLaren, K 2004, 'Bridging the chasm between two cultures', *Skeptical Inquirer*, vol. 28, no. 3, pp. 47–52.

Medvedev, ZA 1978, *Soviet science*, Norton, New York.

Meiland, J 1980, 'What ought we to believe?', *American Philosophical Quarterly*, vol. 17, no. 1, pp. 15–24.

Miller, M 1997, 'Secrets of the cult', *Newsweek*, 129, p. 28.

Mitroff I, Fitzgerald I 1977, 'On the psychology of the Apollo Moon scientists: a chapter in the psychology of science', *Human Relations*, vol. 30, no. 8, pp. 657–73.

Monteleone, TF 1979, 'The gullibility factor', *Omni*, vol. 1, no. 8, p. 146.

Moore, P 1974, *Can you speak Venusian?* David & Charles, Newton Abbot, Devon.

Morphet, C 1977, *Galileo and Copernican astronomy*, Butterworth, London and Boston.

Morris, HM 1984, *Scientific creationism*, Master Books, El Cajon, California.

Myers, FWH 1970, 'High possibilities', in David C. Knight (ed), *The ESP Reader*, Grossett & Dunlop, New York, pp. 421–23.

MythBusters 2003, *MythBusters*, volume 2 (CD), Beyond Properties Ltd.

Nathan, D, Snedeker, M 1995, *Satan's silence*, Basic Books, New York.

National Catholic Reporter 2003, 'Pope considered declaring Mother Teresa blessed, saint in one ceremony', *National Catholic Reporter*, vol. 8, no. 1, p. 39.

National Science Board 2006, *Science and engineering indicators 2004*, <http://www.nsf.gov/statistics/seind04/c3/c3h.htm> (accessed 6 March 2006).

National Science Foundation 2006, *Science and technology: public attitudes and public understanding. Public interest in and knowledge of S&T*, National Science Foundation, 2002, <http://www.nsf.gov/statistics/seind02/c7/c7s1.htm> (accessed 23 February 2006).

Nature, Editorial, 1974, 'Investigating the paranormal', *Nature*, 251, pp. 559–60.

Neimark, J 1996, 'The diva of disclosure', *Psychology Today*, vol. 29, no. 1, pp. 48–55.

New English Bible 1970, *The New English Bible*, New Testament, Oxford and Cambridge University Press, Oxford and Cambridge.

New Zealand Herald 2000, 'Doctors quick to complain on Liam therapy', 9 November 2000.

Numbers, RL 1992, *The Creationists*, Alfred Knopf, New York.

Oakes, ET 1999, 'Pascal: the first modern Christian', *First Things: A Monthly Journal of Religious and Public Life*, August, pp. 41–50.

Oldroyd, D 1980, *Darwinian impacts*, Open University Press, Milton Keynes.

Overton, WR 1982, 'Creationism in schools: the decision in McLean v Arkansas Board of Education', *Science Forum*, 215, pp. 934–43.

Palmer, JNJ 1979, 'The damp stones of positivism: Erich von Däniken and paranormality', *Philosophy of the Social Sciences*, 9, pp. 129–47.

Parejko, K 2003, 'Pliny the Elder: rampant credulist, rational skeptic or both?', *Skeptical Inquirer*, vol. 27, no. 1, pp. 39–42.

Pickering, A 1980, 'The role of interests in high-energy physics', in Karen D Knorr, Roger Krhon and Richard Whitley (eds), *The social process of scientific investigation*, D Reidel, Dordrecht, pp. 107–38.

Plait, P 2002, *Bad astronomy*, John Wiley, New York.

Plato 1953 [nd], *The Dialogues of Plato*, B. Jowett (trans), 4th edition, 4 vols, vol. 1, Oxford University Press, Oxford.

Pliny, the Elder 1962 [nd], *Selections from the history of the world, commonly called the natural history of C. Plinius Secundus*, Philemon Holland (trans), selected by Paul Turner, Southern Illinois University Press, Carbondale, Illinois.

Plummer, M 2004a [1986], 'Psychic surgery – a fraud', in Barry Williams (ed), *The great skeptic CD 2 – in the beginning*, Australian Skeptics 2004, Roseville, NSW, pp. 45–6.

—— 2004b [1986], 'How the MMW got empf'd', in Barry Williams (ed), *The great skeptic CD 2 – in the beginning*, Australian Skeptics 2004, Roseville, NSW, pp. 79–81.

Popper, KR 1968, *Conjectures and refutations: the growth of scientific knowledge*, Harper & Row, New York.

Price, DJ de S 1984, 'The science/technology relationship, the craft of experimental science, and policy for the improvement of high technology innovation', *Research Policy*, vol. 13, no. 1, pp. 3–20.

Puthoff, H, Russell T 1981, 'Rebuttal of criticism of remote viewing experiments', *Nature*, 292, p. 388.

Randi, J 1982, *Flim-Flam!*, Prometheus Books, Amherst, New York.

—— 1983a, 'The project Alpha experiment: part 1. The first two years', *Skeptical Inquirer*, vol. 7, no. 4, pp. 24–33.

—— 1983b, 'The project Alpha Experiment: part 2. Beyond the laboratory', *Skeptical Inquirer*, vol. 8, no. 1, pp. 36–45.

Randles, J, Roberts, A, Clarke, D 2000, *The UFOs that never were*, London House, London.

Rhine, JB 1969, 'Psychical research, or parapsychology', in David C. Knight (ed), *The ESP reader*, Grosset & Dunlap, New York, pp. 2–8.

Riley, JP, Skirrow, G (eds) 1965, *Chemical oceanography*, Academic Press, New York.

Ronson, J 2004, *The men who stare at goats*, Picador, London.

Ross, BL 1978, *Capricorn One*, Futura, London.

Russell, B 1923, *A free man's worship*, Thomas Bird Mosher, Portland, Maine.

—— 1971, *History of Western philosophy*, George Allen & Unwin, London.

—— 1977a (1935) 'Introduction: On the Value of Scepticism', in Bertrand Russell *Sceptical Essays*, pp. 11–21.

—— 1977b [1935], *Sceptical essays*, George Allen & Unwin, London.

Ryan, F 1992, *Tuberculosis: the greatest story never told*, Swift Publishers, Bromsgrove.

Sackett, DL et al 1997, *Evidence-based medicine: how to practice and teach EBM*, Churchill Livingstone, New York.

Sagan, C 1997, *The demon-haunted world. Science as a candle in the dark*, Headline, London.
—— 1997a, 'The fine art of baloney detection', in Carl Sagan (ed), *The demon-haunted world*, pp. 189–206.
Sanders, E 2002, *The family*, Thunder's Mouth Press, New York.
Sceats, R 2000, 'The Benalla dowsing challenge', *Skeptic*, vol. 20, no. 1, pp. 9–10.
Schick, T, Vaughn, L 2002, *How to think about weird things*, McGraw-Hill, New York.
Schmidt, MH et al 1997, 'Does oligoantigenic diet influence hyperactive/conduct-disordered children – a controlled trial', *European Child & Adolescent Psychiatry*, vol. 6, no. 2, pp. 88–95.
Schnabel, J 1993, *Round in circles*, Hamish Hamilton, London.
—— 1997, *Remote viewers: the secret history of America's psychic spies*, Dell, New York.
Schneider, WE 2001, 'Past imperfect', *Yale Law Review*, vol. 110, no. 8, pp. 1531–45.
Shermer, M 1997, *Why people believe weird things*, WH Freeman, New York.
Shermer, M, Grobman, A 2000, *Denying history: who says the Holocaust never happened and why do they say it?*, University of California Press, Los Angeles, California.
Shirer, WL 1964, *The rise and fall of the Third Reich*, Pan, London.
Singer, MT, Lalich, J 1995, *Cults in our midst*, Jossey-Bass, San Francisco.
Singer, P 2001, *Writings on an ethical life*, Fourth Estate, London.
Singh, S 2004, *Big Bang*, Fourth Estate, London.
Sladek, J 1978, *The new Apocrypha*, Granada, London.
Smith, MD, Foster, CL, Stovin, G 1998, 'Intelligence and paranormal belief: examining the role of context', *Journal of Parapsychology*, vol. 62, no. 1, pp. 65–77.
Stanford Encyclopedia of Philosophy 2006, Stanford University, <http://plato.stanford.edu/contents.html> (accessed 27 March 2006).
Stenger, VJ 1990, *Physics and psychics*, Prometheus Books, Amherst, New York.
Stickley, T 2002, 'Parents of baby Caleb found guilty of manslaughter', *New Zealand Herald*, 5 June <http://www.nzherald.co.nz/nz/news/article.cfm?_id=1&objectid=2045066> (accessed 29 June 2009)
Stone, R 2003, 'Championing a 17th century underdog', *Science Forum*, vol. 301, no. 5630, p. 152.
Story, R 1978, *The space gods revealed*, New English Library, London.
Stossel, S 1999, 'Uncontrolled experiment', *New Republic*, vol. 200, no. 13, pp. 17–22.
Sydney Mornhing Herald 2001, 'Nine angry jurors', 6 January 2001, p. 10.
Targ, R, Puthoff, H 1974, 'Information transmission under conditions of sensory shielding', *Nature*, 251, pp. 602–07.
Targ, R, Puthoff, H 1977, *Mind-reach: scientists look at psychic abilities*, Delacorte, New York.
Tart, CT, Puthoff, HE, Targ, R 1980, 'Information transmission in remote viewing experiments', *Nature*, 284, p. 191.
Taylor, AE 1953, *Socrates*, Hyperion, Westport, Conn.
Thalbourne, MA 1995, 'Science versus showmanship: a history of the Randi hoax', *The Journal of the American Society for Psychical Research*, 89, pp. 344–66.
Thalbourne, MA, Delin, PS 1994, 'A common thread underlying belief in the paranormal, creative personality, mystical experience and pychopathology', *The Journal of Parapsychology*, vol. 58, no. 1, pp. 3–39.
Thomas, G, Pring, R (eds) 2004, *Conducting educational research*, Open University Press, Maidenhead, Berks.

Töben, F 2006, The Adelaide Institute (website) 2002, <http://www.adelaideinstitute.org/> (accessed 19 March 2006).

Toulmin, S, Goodfield, J 1966, *The discovery of time*, Harper & Row, New York.

Trefil, JS 1978, 'A consumer's guide to pseudoscience', *Saturday Review*, pp. 16–21.

Trevor-Roper, HR 1969, *The European witch-craze of the sixteenth and seventeenth centuries, and other essays*, Harper & Row, New York.

Tripician, RJ 2000, 'Confessions of a (former) graphologist', *Skeptical Inquirer*, vol. 24, no. 1, pp. 44–7.

Tutt, K 1997, *True life encounters*, Millennium, London.

von Däniken, E 1970, *Chariots of the gods?*, Michael Heron (trans), G. P. Putnam's Sons, New York.

Watson, P 2000, *A terrible beauty*, Weidenfeld & Nicolson, London.

Weber, S 1996, 'It all started with an electronic switch', *Electronic Engineering*, 394, pp. 16–20.

Westrum, R 1978, 'Social intelligence about anomalies: the case of meteorites', *Social Studies of Science*, vol. 8, no. 4, pp. 461–93.

Weyand, RG 2006, 'Why did they bury Darwin in Westminster Abbey?', *Skeptical Inquirer*, vol. 30, no. 1, pp. 33–6.

Whitson, JA, Galinsky, AG 2008, 'Lacking control increases illusory pattern perception', *Science*, 322, 3 October, pp. 115–7.

Williams, B 1993, 'UFO was IPO', *Skeptic*, vol. 13, no. 1, pp. 10–1.

—— 1994, 'Misleading or misreading science', *Skeptic*, vol. 14, no. 4, pp. 51–2.

Wilson, C 1972, *Crash go the chariots*, Word of Truth Publishers, Mount Waverley, Victoria.

Wilson, K, French, C 2006, 'The relationship between susceptibility to false memories, dissociativity and paranormal belief and experience', *Personality and Individual Differences*, 41, pp. 1493–1502.

Wiseman, R, Greening E 1998, 'Psychic exploitation', *Skeptical Inquirer*, vol. 22, no. 1, pp. 50–2.

Wiseman, R, Morris, RL 1995, 'Recalling pseudo-psychic demonstrations', *British Journal of Psychology*, 86, pp. 113–25.

Wiseman, R, Smith, M, Wiseman, J 1995, 'Eyewitness testimony and the paranormal', *Skeptical Inquirer*, vol. 19, no. 6, pp. 29–32.

Woodruff, F 1998, *Secrets of a telephone psychic*, Beyond Words, Hillsboro, Oregon.

Zamulinski, B 2002, 'A re-evaluation of Clifford and his critics', *The Southern Journal of Philosophy*, vol. 40, no. 3, pp. 437–57.

Zanda, B, Rotaru, M (eds) 2001, *Meteorites: their impact on science and history*, Cambridge University Press, Cambridge.

Ziman, JM 1968, *Public knowledge: an essay concerning the social dimension of science*, Cambridge University Press, Cambridge.

Zuckerman, S 1970, *Beyond the ivory tower: the frontiers of public and private science*, Taplinger, New York.

Index

academic skepticism, 68
Aenesidemus, 68
Aetherius Society, 158–9
alternative health care, 2, 5, 48–9, 52, 149
amazing coincidences, 119–24
ancient skeptics, 65–9, 76
anecdotal evidence, 132
anti-Semitism, 173, 178–9
Apollo moon landing, 168
applied science, 20
'Aryan Science', 150–1
astrology, 5
atheism, and skepticism, 106, 108
Australia
  expenditure on alternative and
    complementary health care, 6
  paranormal beliefs, 5
Australian Skeptics, 52, 86
  goals, 90, 98, 100
  opposition to teaching of creation science,
    109
  and use of skepticism, 165

Bacon, Francis, 14, 121–2, 133, 171
Barr, James, 28
Bartz, Wayne R, 97
beliefs
  and formation of scientific knowledge, 25
  impact on results of experiments, 14–15
Bentham, Jeremy, 147
Blackmore, Susan, 157
Blum, Deborah, 105, 107
bodies of knowledge, acceptance and dissent,
  166
Bower, Doug, 159–60
Bratman, Steven, 132–4
Braude, Stephen, 39, 63
breathalyser machines, 166

breatharian cult, 57
Bridgstock, Martin, 4
Britain, paranormal beliefs, 5
burden of proof, 91–2, 105–7
Bush, George W, 117

Capricorn One, 168
Carneades, 68
causality, 78
cereology, 54, 159–60
Chalmers, Alan, 12, 14
Charismatic Christians, 53
Chemical Oceanography, misquoting of, 129–30
Chorley, Dave, 159–60
Christian fundamentalism *see* modern Christian
  fundamentalism
Christianity, rise of, 69
clairvoyance
  impact if true, 7
  testing and ethics, 161–2
Clancy, Susan, 48
client-practitioners, 52–3, 60
Clifford, William Kingdon, 141
Cochrane, Archie, 180
Cochrane Collaboration, 181
cold fusion, 118
Cole, Stephen, 18–19
Committee for the Scientific Investigation of
  Claims of the Paranormal (CSICOP)
  aims, 89–90, 98
  foundation of, 86, 88
  name change, 109
Committee for Skeptical Investigation
  foundation, 109, 153
  goals, 109, 153, 165
consequentialism, 140–1, 147–8, 153, 155, 158
conspiracy theories, 168, 173
continental drift, 22–3

197

Printed in the United States
By Bookmasters